OFDM-Based Broadband Wireless Networks

OFDM-Based Broadband Wireless Networks

Design and Optimization

Hui Liu

Guoqing Li

A JOHN WILEY & SONS, INC., PUBLICATION

Library of Congress Cataloging-in-Publication Data:

Liu, Hui, 1968–
    OFDM-based broadband wireless networks : design and optimization / Hui Liu, Guoqing Li.
        p. cm.
    Includes bibliographical references and index.
    ISBN-13 978-0-471-72346-2 (cloth)
    ISBN-10 0-471-72346-0 (cloth)
      1. Orthogonal frequency division multiplexing.   2. Wireless communication systems.
      I. Li, Guoqing, 1974–   II. Title.
    TK5103.484.L58 2005
    621.382'16—dc22

                                                                        2005011809

Printed in the United States of America.

10 9 8 7 6 5 4 3 2 1

To our families

Hui Liu
Guoqing Li

Contents

x

Preface

Emerging technologies such as WiFi and WiMAX are profoundly changing the landscape of wireless broadband. As we evolve into future generation wireless networks, a primary challenge is the support of high data rate, integrated multimedia type traffic over a unified platform. Due to its inherent advantages in high-speed communication, orthogonal frequency division multiplexing (OFDM) has become the modem of choice for a number of high profile wireless systems (e.g., DVB-T, WiFi, WiMAX, Ultra-wideband).

The book aims at providing wireless professionals and graduate students an up-to-date treatment of the subject area and, more importantly, the technical concepts which are at the core of broadband air-interface design and implementation. Our goal was to produce a textbook which would provide enough background material and discuss advanced principles that could enable significant improvements in network characteristics not realizable with current wireless infrastructure.

For readers interested in the WiFi and WiMAX standards, an appendix describing the latest innovations and applications specifically related to these standards is provided. On the other hand, the technical discussion in this book is not narrowly focused on any specific standard. Instead, each chapter contains a clear exposition of the fundamental aspects of the topic. An important thread in this book emphasizes design concepts and algorithms for the air-interface of OFDM-based broadband wireless access networks. We are interested in protocols that can capture the full potential of OFDM by jointly optimizing the link-level and the system-level performance metrics. The technical audience will be exposed to modern principles and methodologies beyond the current wireless design paradigm. The coverage includes established techniques as well as an ensemble of research results and articles by the authors that deal with OFDM modem and OFDMA-based multiple-access schemes. This mix should be beneficial not only to entry level students needing a comprehensive understanding of OFDM, but also to senior graduate students and practicing engineers seeking wireless system design and optimization guidelines.

Chapters 1 and 2 take a comprehensive look at OFDM, its history, principles and applications. Design challenges arising from broadband fading channels and multimedia traffic are discussed. These chapters should have appeal to a broad audience in understanding the latest trends in broadband wireless technologies.

Chapters 3 and 4 are tailored for physical layer researchers and modem design engineers. Chapter 3 in particular covers the various types of enabling techniques for OFDM modem, while Chapter 4 provides an extensive treatment on MIMO and smart antennas and their integration with OFDM. Design methodologies that are in use or being proposed for future generation systems (e.g., WiMAX and WiFi) are described.

Chapters 5, 6 and 7 deal with the MAC functionalities and present some system considerations for OFDMA-based cellular networks. These chapters cover the important problem of radio resource management through multiple access control, cross-layer optimization and frequency planning. Issues relevant to "multiuser diversity" are treated both from information theoretical and system protocol standpoints.

Appendices which cover IEEE 802.11a/g and IEEE 802.16e provide details for those interested in the latest development in WiFi and WiMAX standards.

Readers are also referred to ftp://ftp.wiley.com/public/sci_tech_med/ofdm for future updates regarding this book.

We would like to acknowledge all those who have contributed to the preparation of this book. Many colleagues and students, past and present, have contributed their ideas. The contributions of Dr. Guanbin Xin, Dr. Manyuan Shen, and Prof. Uf Tureli, are particularly noteworthy. The performance analysis on WiMAX system by Wolf Mack at Adaptix Inc. and the detailed review by Anatoliy Ioffe have been highly useful for improving this book. It is also a pleasure to acknowledge those who attended the "Mobile Broadband Network" course at the University of Washington for their helpful suggestions and corrections. Finally, we would like to thank our home institute, the University of Washington, and NSF and ONR for supporting our research activities in wireless OFDM network over the past years.

Chapter 1

Introduction

The wireless industry is undergoing a major evolution from narrow-band, circuit-switched legacy systems to broadband, IP-centric platforms. A common theme in this broadband evolution is the use of OFDM modem and open network architectures. This chapter discusses the modern view of packet-based broadband wireless communications and their associated challenges. We will cover some of the latest technological advances in digital broadcasting, wireless local area network (LAN), and beyond 3rd generation mobile networks. The roles of OFDM modem and MAC protocols in an air interface are described. These modules must be synergistically integrated in order for the network to (1) achieve high spectral efficiency per unit area (bit/s/Hz/square-meter), and (2) to meet the anticipated peak throughput requirements for multimedia traffic.

1.1 OFDM-based wireless network overview

OFDM has become one of the most exciting developments in the area of modern broadband wireless networks. Although the notion of multicarrier transmission or multiplexing (e.g., frequency-division multiplexing - FDM) can be dated back to 1950s, high spectral efficiency and low cost implementation of FDM became possible in the 1970s and 1980s with advances in Digital Fourier Transform (DFT). It is not until the 1990s that we witnessed the first commercial OFDM-based wireless system – the digital audio broadcasting (DAB) standard. A few historical notes are listed below

- 1958: Kineplex, a military multi-carrier high-frequency communication system [1];

- 1966: R. W. Chang at Bell labs describes the concept of using parallel data transmission and FDM [2];

- 1970: First patent issued on OFDM [3];

- 1971: Weinstein & Ebert, and later Hirosaki in 1981, proposed DFT implementation of FDM [4];

- 1995: ANSI standard T1.413: discrete multitone modulation part of the ADSL standard [6];

- 1995 ETSI DAB standard: first OFDM-based wireless standard for digital audio broadcasting [7];

- 1997: DVB-T: terrestrial digital video broadcasting standard [5];

- 1999: IEEE 802.11a and HIPERLAN/2 standard for wireless LAN;

- 2004: IEEE 802.16a/d standard for fixed broadband wireless MAN [6];

- 2005: OFDM-based mobile cellular networks being developed under IEEE 802.16e and IEEE 802.20.

The most prominent feature of OFDM-based systems is the high data rate. Below we give an overview of three systems that are either widely deployed or beginning to change the landscape of wireless communications.

1.1.1   Digital broadcasting and DVB-T

When first introduced, television programs were analog signals distributed wirelessly through broadcasting. DVB (Digital Video Broadcasting) is a consortium of around 300 companies from more than 35 countries, in the fields of broadcasting, manufacturing, network operation and regulatory matters that have come together to establish common international standards for the move from analog to digital broadcasting – see URL: `Http://www.dvb.org`. DVB-T in particular, provides terrestrial digital video broadcasting services in the VHF (130-260 MHz) and the UHF (480 - 862 MHz) bands. In May 1998, a consortium of 17 broadcasters, network operators, manufacturers of professional and domestic equipment, and research centers launched the MOTIVATE project. MOTIVATE has investigated the practical and theoretical performance limits of DVB-T for mobile reception.

DVB-T makes efficient use of frequency and could carry data, speech and Internet pages as well as TV in the MPEG-2 transport stream. Table 1.1 lists the data rates of different audio and video streams. The basic requirements of a DVB-T service include

- large capacity, high data rate (SDTV, EDTV, and HDTV)

- single-frequency terrestrial networks

Quality	Data Rate
telephone	64 Kbps
CD	1.4Mbps
DVD	3.5 - 6 Mbps
MPEG-2 SDTV	5 Mbps
MPEG-2 EDTV	10 Mbps
MPEG-2 HDTV	20-30 Mbps

Table 1.1: Data rate of multimedia services

- optimum coverage for stationary reception with a rooftop antenna; support portable, but not mobile reception

- simultaneous broadcasting of low rate (robust) steams and high rate (fragile) streams

DVB-T addresses the above needs with an OFDM modem and a number of technological innovations. In particular, DVB-T employs two variants of the coded OFDM (COFDM) technique with either a size 2K-FFT or a size 8K-FFT. When combined with QPSK, 16-QAM, and 64-QAM, DVB-T is capable of delivering a peak rate of 31.67 Mbits/s, although the most common operation mode (QPSK + 1/2 coding) offers around 5Mbit/s throughput. The key enabling technologies in DVB-T are:

- coded OFDM modem

- hierarchical modulation

- continuous and distributed pilots

The significance of these techniques will be explained in Chapters 2 and 3 in which we provide an in-depth treatment on COFDM and pilot designs. Since 2000, DVB-T has been modified to provide other wireless broadcasting applications such as the DVB-H (for handheld terminals). Continuous evolution in technology and services is anticipated in DVB-T. There is even a possibility of converging DVB-T with telecommunication networks such as the 3G and WiMAX.

1.1.2 Wireless LAN and IEEE 802.11

As of today, the wireless LAN is arguably the most popular broadband wireless network in the world. Standard LAN protocols, such as the Ethernet, that operate at fairly high speeds with inexpensive connection hardware can bring digital networking to almost any computer. The IEEE standard for wireless

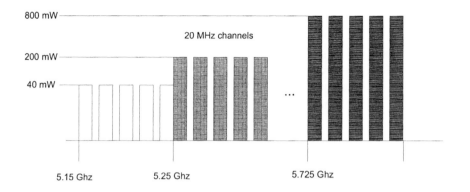

Figure 1.1: Maximum emission in the UNII band

LANs (IEEE 802.11) introduces *mobility and flexibility* to a LAN environment by allowing computers and other devices to communicate with one another wirelessly (URL: Http://grouper.ieee.org/groups/802/11/). This technology is beneficial for improved access to fixed LAN and inter-network infrastructure (including access to other wireless LANs) via a network of access points, as well as creation of higher performance ad hoc networks.

The two wireless LAN solutions that utilize OFDM are the 802.11g and the 802.11a. The IEEE's 802.11g standard is designed as a higher-bandwidth (54 Mbps) successor to the popular 802.11b, or the Wi-Fi standard (11 Mbps). While 802.11g is backwards compatible with 802.11b at the 2.4GHz ISM (industrial, scientific & medical) band, the 802.11a operates in the newly allocated UNII band (Unlicensed National Information Infrastructure). Figure 1.1 shows the channel plan and the associated maximum emission of the UNII band. Because of the power limitation, most applications of wireless LAN are limited to homes and office buildings.

The basic requirements of wireless LAN include

- single MAC to support multiple PHYs

- overlap of multiple networks

- robustness to interference

- mechanisms to deal with "hidden nodes"

- provisions for time-bounded services

To meet these requirements, a number of challenges must be addressed. The 802.11a/g utilizes an OFDM modem to deliver a range of data rates from 6 up to 54 Mbps. Even higher speed versions (802.11n) are being ratified to blast data

Data rate	6, 9, 12, 18, 24, 36, 48, 54Mbps
Modulation	BPSK, QPSK, 16-QAM, 64-QAM
Coding rate	1.2, 2/3, 3/4
Number of subcarriers	52
Number of pilots	4
OFDM symbol duration	4us
Guard interval	800ns
Subcarrier spacing	312.5kHz
-3 dB bandwidth	16.56MHz
channel spacing	20Mhz

Table 1.2: The 802.11a/g modem parameters

rate beyond 100 Mbps. The parameters of its OFDM modem are summarized in Table 1.2.

At the current stage, the major limitations of the wireless LAN are its coverage (several hundred feet) and quality of service (QoS) support. While the coverage issue is more regulatory than technical, the lack of QoS is inherent due to 802.11's contention based MAC. Enhancements to the current 802.11 MAC are ongoing to expand support for LAN applications with Quality of Service requirements. The 802.11e is a work group to provide improvements in security and protocol capabilities and efficiency. These enhancements, in combination with recent improvements in PHY capabilities from 802.11a and 802.11b, will increase overall system performance and expand the application space for 802.11. Example applications include transport of voice, audio and video over 802.11 wireless networks, video conferencing, media stream distribution, enhanced security applications, and mobile and nomadic access applications. More discussions on 802.11 can be found in the appendix chapter of this text.

1.1.3 WiMAX and IEEE 802.16

Another OFDM-based broadband wireless data solution that has gained broader industry acceptance is the IEEE 802.16. This emerging standard complements 802.11 as a truly metropolitan area network (MAN). With both licensed and license-exempt options, the typical applications of 802.16 include mesh networks, backhaul, wireless DSL to residential and small-office-and-home-office (SoHo), and broadband mobile networks. The peak data rate of 802.16 can achieve 70Mbps, with coverage up to 35 Km in a line-of-sight, fixed environment.

The IEEE 802.16 standard defines a medium access control (MAC) networking layer that supports a number of physical layer specifications. The initial

802.16 standard was followed by several working groups, some of whom have released their amendments to the standard. The most prominent amendments are the 802.16a, which extends the standard into the spectrum between 2 and 11 GHz; the 802.16d, which defines the system profiles for 802.16a implementation; and the 802.16e, which is in development to add mobility to stations that primarily support fixed wireless networking in the 2 to 6 GHz bands. Dubbed as *mobile WiMAX*, the 802.16e can potentially rival the code division multiple access (CDMA)-based 3G paths in large-scale deployment.

Other than the coverage and mobility, a key difference between the 802.11 and the 802.16 is the MAC. Unlike the 802.11, which supports 10's of users, the 802.16 MAC is designed to support thousands of users using a grant-request mechanism. The QoS support for voice and video is designed from ground up, and differentiated service levels are also introduced. The security features of 802.16 are also much more advanced. Technologically, the 802.16 encompasses some of the most cutting-edge wireless innovations including

- scalable OFDMA

- variable channel configurations for multiuser diversity exploitation

- multiple-input multiple-output (MIMO) and advanced antenna systems

- advanced channel coding and hybrid-ARQ

- QoS and service classes.

More discussions on 802.16e can be found in the Appendix chapter of this text.

1.2   The need for "cross-layer" design

Broadly speaking, the design of a wireless system involves link-level issues and system-level issues. The wireless link-level primarily addresses two challenges that arise from the physical medium - namely, the channel fading (potentially both time and frequency selective) and the multiple-access interference. Advances in link design for wireless channels have led to modulation schemes and channel coding schemes that provide increased robustness to MAI and multipath and thereby, enhance link level or radio capacity. OFDM, in particular, is arguably the most significant one in the current broadband evolution given its dominant position in almost all broadband systems. Its form of modulation that transmits high-speed data via multiple parallel low-rate streams presents excellent performance over frequency selective and interfered channels.

While OFDM provides a powerful physical layer engine for broadband communications, applying it without thorough *system* considerations may lead to disappointing results or even negatively impact the overall performance. For

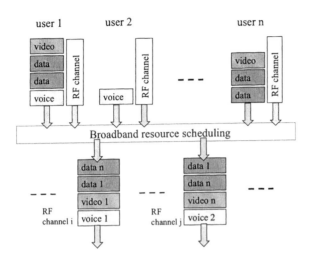

Figure 1.2: A channel- and application-aware MAC for dynamic radio resource allocation

example, if one simply combines OFDM with TDMA (time-division multiple-access) to provide broadband services in a cellular type of environment, the system will run into the coverage problem due to the link budget bottleneck[1]. In addition, the coarse granularity of OFDM/TDMA will severely hinder the QoS offerings to a large number of subscribers, which inevitably reduces the radio resource efficiency. We will show in Chapter 4 that a judiciously designed multiple-access scheme not only can retain all the benefits of OFDM, but also capture additional gains unavailable to traditional CDMA systems. The basic idea is illustrated in Figure 1.2 where the radio allocation module is designed to be both channel-aware and application-aware through cross-layer interactions.

In the Appendix, the application scenarios of broadband wireless networks such as WiFi and WiMAX will be described. A common design objective of these types of broadband systems is to support heterogeneous traffic with different QoS requirements on unified network architecture. For example, voice is significantly lower rate than video and has low delay tolerance; video/data are higher rate and can withstand greater delay and delay variability. In order to deliver the many types of services, at least the following has to be taken into account in the design and evaluation of the air interfaces:

- *Data rate per user/link*: The peak data rate per user supported by the

---

[1]In TDMA, each subscriber must burst over the entire channel in its given time slots, resulting in very high peak rate.

system has to be high enough to support high-speed applications (e.g., the HDTV with data rate on the order of 10 Mb/s to 100 Mb/s).

- *Granularity*: The rich-media services of the future will have rate requirements of large variance from individual subscribers (e.g., from 10kbps voice to 1.5Mbps MPEG video). In order to prevent unnecessary overhead, it is essential that the air interfaces possess the finest granularity.

- *QoS and service classes*: the system must be able to efficiently support a wide range of applications with diverse QoS parameters. In addition, the wireless network should be able to accommodate different service classes (each of which identifies a specific set of QoS parameters) by configuring its service flows.

- *System capacity*: The capacity and number of users per cell are of importance for the system design and strongly depend on the chosen multiple access schemes and the frequency re-use factor. The trade-off between maximizing capacity and coverage will be a complex design topic.

- *Simplicity and scalability:* Ideally, the interfaces between different network components should be open (e.g., IP-based), allowing standard modules to interact without jeopardizing the integrity of the entire systems. In addition, the network should be easily scalable based on a future-proof architecture.

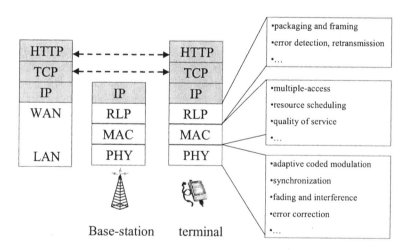

Figure 1.3: The basic layers of a broadband network

These requirements inevitably lead to a long list of design criteria. The list of competing techniques for addressing them is much longer. It is therefore important to prioritize the complex design criteria. The most intriguing part of engineering design is to understand the impact and tradeoffs of individual techniques from a system viewpoint [10]. By synergistically integrating them, impact of individual techniques on overall network performance may be amplified without undue complexity. Figure 1.3 illustrates a layered model of a wireless air-interface and their essential functions. Entering the wireless side, the external network data (e.g., IP packets) are first handled by radio link protocol (RLP) layer which packs (and unpacks) them into wireless protocol data units. The medium access-control (MAC) layer provides core functionality of system access and resource allocation, while the physical (PHY) layer handles the payload over the wireless medium. The MAC also performs connection establishment and connection maintenance etc. Traditionally, the air-interface protocols have managed network access and congestion via appropriate multiple access and routing protocols. Much of the effort in designing the MAC/RLP layer on top of the PHY seems to have followed (with only some notable recent exceptions) the mode of 'adapting' ideas from protocols designed primarily for wired networks. Clearly, these are inadequate (or at best seriously sub-optimal) approaches for broadband that puts a premium on achieving QoS specifications in increasingly demanding scenarios. Meeting these challenges will require technical innovation at both the networking and the physical layers.

Design optimization of OFDM for cellular applications has been identified by many research organizations as the key tasks of future broadband networks. The proper utilization of OFDM to optimize the network properties for beyond 3rd generation (B3G) and 4th generation (4G) has just started. In light of the complexity and robustness issues of such protocols, our attempt in the ensuing chapters is to develop suitable models and methodologies for the rigorous design of high performance air-interface taking into account the network dynamics. The objective of this book is not to produce an actual system. Instead, the approach is to first provide the readers with detailed knowledge of the essential blocks of OFDM. Following that, important insights regarding how to apply these techniques in a top-down system design are presented. With an established system approach, it will become more convenient to produce the specifications of wireless broadband protocols much beyond the scope of current platforms.

1.3 Organization of this text

This book consists of 8 chapters that provide a comprehensive coverage on the PHY and MAC layers of OFDM-based broadband wireless air interface. Chapters 1 and 2 describe the general framework of OFDM, Chapters 3 and 4 discuss enabling techniques pertaining to the physical layer, and Chapters 5 to 7 offer the design guidelines from a system/network prospective. Specifically,

- Chapter 1 overviews OFDM applications in broadband wireless networks. Design criteria for modern broadband multimedia systems are established, leading to the notion of *cross-layer* optimization.

- Chapter 2 characterizes broadband fading channels and describes the basic principles of OFDM modem. By showing how multipath fading can be effectively mitigated with parallel narrowband channels, the chapter establishes OFDM as the canonical form for broadband modem. Transceiver diagrams of OFDM are presented, as well as design considerations on practical OFDM systems.

- Chapter 3 deals with the signal processing aspects of OFDM communications. To cope with impairments from fading channels and hardware imperfections, a set of tasks such as frequency synchronization, channel estimation, and phase noise compensation, must be performed before OFDM demodulation. This chapter describes a number of highly effective algorithms that can significantly improve the performance of an OFDM system. It contains contributions from Dr. Guanbin Xin on phase noise estimation and I/Q imbalance compensation.

- Chapter 4 discusses spatial processing in OFDM applications. By first laying the groundwork on antenna array operations, the chapter quantifies the potential of space-time processing with the information theoretic capacity of a multiple-input multiple output (MIMO) system. Examples of space-time codes are presented, together with schemes suitable for practical OFDM systems. Part of the chapter is based on the contributions of Dr. Manyuan Shen on outdoor space-time beamforming for OFDM.

- Chapter 5 considers the MAC layer of an OFDM-based network and, in particular, the orthogonal frequency-division multiple-access (OFDMA) scheme. The topic of focus is *multiuser diversity* and, specifically, how it can be exploited within OFDMA through dynamic resource allocation. The chapter answers a fundamental question regarding the optimality of OFDMA in both SISO and MIMO configurations. In addition, the air interface protocol needed to capture the multiuser diversity gain is described.

- Chapter 6 discusses the design considerations in practical OFDMA systems. In particular, a cross-layer design issue pertaining to the OFDMA traffic channel configuration is addressed. The optimal configuration for fixed/portable and mobile users are derived, respectively. This chapter also describes the scalable OFDMA design in IEEE 802.16e. The results provide some important guidelines for OFDMA system engineers.

- Chapter 7 addresses radio resource allocation in a multicell OFDMA network. Built upon the concept of dynamic channel allocation (DCA) in narrowband networks, the chapter presents centralized and distributed DCA approaches for broadband systems. The results reveal essential tradeoffs regarding the network spectral efficiency as a function of the cell/sector configuration.

- Chapter 8 is a tutorial on the IEEE 802.11a/g standard and the IEEE 802.16d/e standard. Commonly known as the WiFi and the WiMAX, these two OFDM-based broadband wireless systems have gained increased popularity in the past years. The chapter summarizes the most important features in both standards while presenting insights regarding their implementation and applications. References to advanced techniques described throughout this book are provided, along with discussions on how they can benefit the performance of the WiFi and WiMAX networks.

Bibliography

[1] R. R. Mosier and R. G. Clabaugh, "Kineplex, a bandwidth efficient binary transmission system," *AIEE Trans.*, vol. 76, pp. 723-728, January 1958

[2] R. W. Chang, "Synthesis of band limited orthogonal signals for multichannel data transmission," *Bell Syst. Tech. Journal*, vol. 45, pp. 1775-1796, Dec. 1966

[3] "Orthogonal frequency division multiplexing," U.S. Patent no. 3,488,4555, 1970

[4] S. B. Weinstein and P. M. Ebert, "Data transmission by frequency division multiplexing using the discrete Fourier transform," *IEEE Trans. Communications*, vol. COM-19, pp. 628-634, Oct. 1971

[5] B. Hirosaki, "An orthogonally multiplexed QAM system using the discrete Fourier transform," *IEEE Trans. Comm.*, vol. COM-29, pp. 982-989, July 1981

[6] "Network and Customer Installation Interfaces — Asymmetric Digital Subscriber Line (ADSL) Metallic Interface," ANSI standard T1.413, 1995

[7] ETS 300 401, "Digital audio broadcasting (DAB); DAB to mobile, portable and fixed receivers,"

[8] ETS 300 744, "Digital video broadcasting; framing. structure, channel coding and modulation for digital terrestrial television (DVB-T)." European Telecommunications Standards Institute ETSI, January 1999

[9] IEEE P902.16-2004, "standard for local and metropolitan area networks Part 16: air interface for fixed broadband wireless access systems;" grouper.ieee.org/groups/802/16/tgd/

[10] David J. Goodman, Wireless Personal Communications Systems, Addison-Wesley Wireless Communications Series, 1997

Chapter 2

OFDM Fundamentals

Despite being a nearly 50-year-old concept, it is only in the last decade that OFDM becomes the modem of choice in wireless applications. One of the biggest advantages of an OFDM modem is the ability to convert dispersive broadband channels into parallel narrowband subchannels, thus significantly simplifying equalization at the receiver end. Another intrinsic feature of OFDM is its flexibility in allocating power and rate optimally among narrowband sub-carriers. This ability is particularly important for broadband wireless where multipath channels are "frequency-selective" (due to cancellation of primary and echoed signals). From a theoretical standpoint, OFDM was known to closely approximate the "water-filling" solutions of information theory that are capacity achieving. Some early work of Weinstein and Ebert [7] and Hirosaki [8] based on an FFT implementation of OFDM achieved both complexity and decoded bit count that was comparable to single-carrier counterparts. OFDM potential came to fruition in the designs of discrete multi-tone systems (DMT) for xDSL/ADSL applications, IEEE 802.11.a wireless LAN, digital broadcasting systems DAB-T/DVB-T, the recent 802.16 broadband wireless access. A highlight of the wireless OFDM landscape is depicted in Figure 2.1.

This chapter describes the underlying principles of OFDM modem. To better appreciate the capability of OFDM in combating channel impairments, we will first devote the space to characterize the wireless fading phenomenon. By identifying the types of signal distortions a channel may cause, the choice of OFDM as a broadband modem solution will become evident.

2.1 Broadband radio channel characteristics

Two difficulties arise when a signal is transmitted over the wireless medium. The first is *envelope fading*, which attenuates the signal strength in an unpredictable way. The other is *dispersion*, which alters the original signal waveforms in both time and frequency domains.

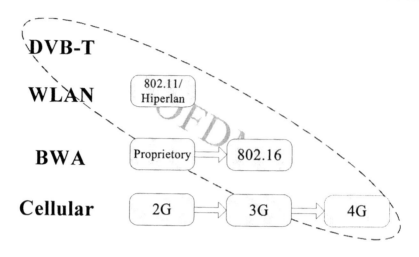

Figure 2.1: OFDM in broadband wireless networks

2.1.1 Envelope fading

The envelope fading manifests itself in the form of fluctuations in amplitude of received signals. The main causes are multipath reflections. Consider a scenario where the transmitted signal arrives at the receiver through two paths with negligible delay between them. The random scattering gives rise to different path attenuations in α_1, α_2.

$$x(t) = \alpha_1 s(t) + \alpha_2 s(t) = (\alpha_1 + \alpha_2)s(t).$$

In this case, the channel response can be modelled as a single delta function with a random envelope. Assuming α_1, α_2 are equal strength complex Gaussian, then the envelope of their sum, $r = |\alpha_1 + \alpha_2|$, obeys a *Rayleigh* distribution:

$$p(r) = \frac{r}{\sigma^2} e^{-\frac{r^2}{2\sigma^2}}$$

with the mean value and variance

$$E\{r\} = \sigma\sqrt{\tfrac{\pi}{2}}, \quad \sigma_r^2 = \sigma^2\left(\tfrac{4-\pi}{2}\right)$$

In the case when the multipath components are not of the same strength (e.g., dominant line-of-sight scenarios), the envelope can be more accurately described by the *Rice* distribution [2].

In addition to the signal strength, the wireless medium may also affect the original signal through dispersion, which includes time dispersion (frequency selective) and frequency dispersion (time selective) fading.

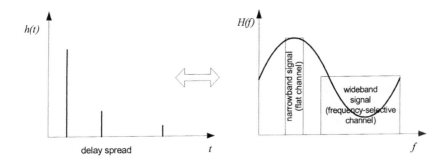

Figure 2.2: A time dispersive (frequency-selective) channel and its effect on narrow- and broad-band signals

2.1.2 Time dispersive channel

The arrival time of scattered multipath signals are inevitably distinct. Whether these delays smear the transmitted signal depends on the product of the signal bandwidth and the maximum differential delay spread. A pictorial view of the time dispersive channel is depicted in Figure 2.2.

The multipath channel can be represented as a linear transfer function $h(t)$. Because of the different propagation delays, the channel impulse response is superposition of delayed delta functions:

$$h(t) = \sum_{m=0}^{M-1} \alpha_i \delta(t - \tau_m)$$

In the case of Figure 2.2, $M = 2$.

Since the multipath delays, $\{\tau_m\}$, are distinct, the frequency response of $H(f) = \mathcal{F}\{h(t)\}$ will exhibit amplitude fluctuation. Such fluctuation in the frequency domain will distort the waveform of a broadband signal. More specifically in digital communication, a channel is considered *frequency-selective* if the multipath delays are distinguishable relative to the symbol period T_{symbol}:

$$\tau_{\max} - \tau_{\min} \approx T_{symbol} = \frac{1}{\text{BW of signal}} \Leftrightarrow (\tau_{\max} - \tau_{\min}) \times (\text{BW of signal}) \approx 1$$

On the other hand, if the signal bandwidth is sufficiently narrow, the channel frequency response within the signal bandwidth can be approximated as constant. A wireless channel is considered *flat* if the multipath delays are indistinguishable relative to the symbol period:

$$\tau_{\max} - \tau_{\min} \ll T_{symbol} = \frac{1}{\text{BW of signal}} \Leftrightarrow (\tau_{\max} - \tau_{\min}) \times (\text{BW of signal}) \ll 1$$

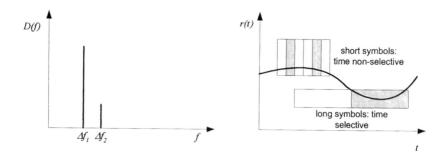

Figure 2.3: A frequency dispersive (time-selective) channel and its effect on short and long symbols

The often used parameters in characterizing a time dispersive channel include [2]:

- mean excess delay

- root-mean squared (rms) delay spread: τ_{rms}

- excess delay spread

- coherence bandwidth: B_c.

2.1.3 Frequency dispersive channel

The short-term fluctuation of the received signal in time domain can be best explained by the Doppler effects due to movement of the transmitter, the receiver, or the environment. The Doppler effect is multiplicative in time, rendering the channel impulse response linear, but time variant.

Consider Figure 2.3 which depicts the Doppler shifts associated with two multipaths in the frequency domain. For simplicity, let us assume the delay spread between the two multipath signals is negligible. At the baseband, the received signal is given by

$$
\begin{aligned}
x(t) &= s(t)e^{j2\pi\Delta f_1 t} + \alpha s(t)e^{j2\pi\Delta f_2 t} \\
&= \left(e^{j2\pi\Delta f_1 t} + \alpha e^{j2\pi\Delta f_2 t}\right)s(t).
\end{aligned}
$$

The Dopplers introduce two types of distortion effects to the received signals: (i) signal variation over time, and (ii) broadened signal spectrum. Define channel coherence time as

$$
T_c = 1/\Delta f_{\max}
$$

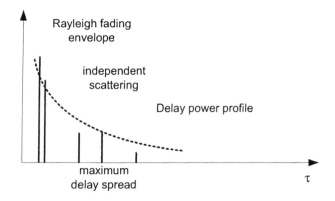

Figure 2.4: A time and frequency dispersive channel profile

where Δf_{max} is the maximum Doppler frequency. When the Doppler shift is comparable to the signal bandwidth (i.e., coherence time $T_c \sim$ the symbol period), the channel is termed *time selective* (fast fading) or *frequency dispersive*. On the other hand, if the Doppler shift is insignificant relative to the symbol rate (channel coherence time $\gg$ symbol period), the channel is termed time nonselective (slow fading).

2.1.4 Statistical characteristics of broadband channels

In reality, a wireless channel may be both time dispersive and frequency dispersive at the same time. Given its random nature, a system design must be based upon the statistical characteristics of wireless channels.

Mathematically, the time and frequency dispersive channel can be modeled as a linear time-variant (LTV) transfer function. A commonly used model for a broadband channel in Rayleigh fading assumes

- M significant solvable uncorrelated paths with normalized delays (by symbol duration T_{symbol}): $\tau_0, \tau_1, ..., \tau_{M-1}$ ($\tau_0 = 0$).

- the M path gains are complex Gaussian random variables $\{\alpha_0, \alpha_2, ..., \alpha_{M-1}\}$, having independent real and imaginary parts with zero mean and variance $\sigma_i^2/2$.

A profile of the LTV channel is depicted in Figure 2.4.

Assuming f_c is the carrier frequency, the time-variant channel impulse response is given by

$$h(\tau, t) = \sum_{m=0}^{M-1} \alpha_i e^{-j2\pi f_c \tau_m(t)} \delta(\tau - \tau_m(t))$$

Clearly, $h(\tau, t)$ is a complex-valued Gaussian random process in the t variable. Its envelope at any instant t is therefore Rayleigh-distributed. The time-varying $\tau_m(t)$ in $e^{-j2\pi f_c \tau_m(t)}$ captures the spectrum-smearing Doppler effect.

Applying the Fourier transform with respect to the delay variable τ, we obtain the *time-frequency channel response* of the time-variant channel:

$$h(f, t) = \sum_{m=0}^{M-1} \alpha_i e^{-j(\theta_m + 2\pi F_D t - 2\pi f \tau_m)} \qquad (2.1)$$

Here, we re-express $2\pi f_c \tau_m(t)$ as $(2\pi F_D t + \theta_m)$ where F_D is the Doppler frequency and θ_m is a random phase attribute to the mth multipath.

In most analysis, the wide-sense stationary uncorrelated scattering (WSSUS) channel model is used [1]. The statistical characteristics are then described by its covariance matrix

$$R_h(\Delta f, \Delta t) = E\{h(f; t)h^*(f - \Delta f; t - \Delta t)\} \qquad (2.2)$$

Several important profiles can be derived from the autocorrelation functions:

- *The frequency correlation function*: $p_h(\Delta f) = R_h(\Delta f, 0)$, quantifies the channel correlation in frequency domain. The nominal width of $p_h(\Delta f)$, termed *coherence bandwidth* B_C, is a statistical measure of the range of frequencies over which the channel can be considered flat.

- *The delay power profile*: $p_h(\tau) = \mathcal{F}^{-1}\{p_h(\Delta f)\}$, quantifies the time dispersive properties of the channel. The nominal width of $p_h(\tau)$ is known as the *multipath delay spread* $\tau_{\max}$. The rms delay spread τ_{rms} is defined as the square root of the second central moment of the delay power profile. In practice, we often use the following approximation

$$B_C = \frac{1}{5\tau_{rms}} \qquad (2.3)$$

- *The time correlation function*: $p_h(\Delta t) = R_h(0, \Delta t)$ quantifies the time varying nature of the channel. Its Fourier transform is the *Doppler power spectrum* $\Phi_h(v) = \mathcal{F}\{p_h(\Delta t)\}$. The nominal width of $\Phi_h(v)$, termed the *Doppler spread* B_D, is defined as the range of frequencies over which the Doppler spectrum is essentially non-zero. The inverse of the Doppler spread,

$$T_C = 1/B_D \qquad (2.4)$$

is defined as the *channel coherence time*, which is a statistical measure of the time interval over which the channel response is essentially invariant.

To obtain Rayleigh fading with the Jakes' spectrum and an exponentially decaying power delay profile with RMS-value τ_{rms}, the following probability density functions are chosen for Eq (2.1):

$$p_\theta(\theta) = 1/2\pi$$

$$p_F(F_D) = \frac{1}{\pi F_{D,\max} \sqrt{1-(F_D/F_{D,\max})^2}}$$

$$p_\tau(\tau) = \frac{e^{-\tau/\tau_{rms}}}{\tau_{rms} \left(1 - \tau_{\max}/\tau_{rms}\right)}$$

In summary,

- Envelope fading affects the signal strength and therefore fading margin in link budget calculation of the wireless system. Power control and spatial diversity techniques are among the most effective means to cope with envelope fading;

- Frequency-selective fading alters the signal waveform and therefore the detection performance. Traditionally, channel equalization is utilized to compensate the effect. Alternatively, one can overcome the frequency selectivity by transferring a broadband signal into parallel narrowband streams as shown in ensuing discussion;

- Time-selective fading smears the signal spectrum and introduces variation too fast for power control. Time interleaving and diversity techniques are most effective means of coping with time-selective fading.

2.2 Canonical form of broadband transmission

OFDM is a form of multicarrier modulation that transmits broadband data over parallel narrowband streams. The superiority of OFDM over single-carrier based modems can be better understood by answers to the following three fundamental questions:

1. Q1: *What kind of signal waveforms are immune to multipath effects?*

2. Q2: *How tight can we pack signals in a given bandwidth?*

3. Q3: *Does fast implementation exist for broadband modem?*

Q1: *What kind of signal waveforms are immune to multipath?*

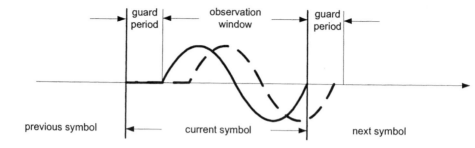

Figure 2.5: Zero-padding guard interval

Multipath induced intersymbol interference is traditionally handled with equalizers at the receiver side. On the other hand, if the signal waveform is designed to be immune (i.e., to a scalar ambiguity) to multipath distortions, the receipt can be dramatically simplified. Since the multipath channel can be perfectly modeled as a linear system, the right signal waveforms can be derived by examining the input-output relation of a linear operation:

$$y(t) = h(t) * x(t) = \int_{-\infty}^{\infty} h(\tau) x(t - \tau) d\tau$$

It is well known that with a complex exponential input, $x(t) = s \cdot w(t) = se^{j\omega t}$, the output signal of the linear channel is identical to the input (within a multiplicative constant):

$$
\begin{aligned}
y(t) &= h(t) * x(t) \\
&= s \int_{-\infty}^{\infty} h(\tau) e^{-j\omega(t-\tau)} d\tau \\
&= se^{-j\omega t} \int_{-\infty}^{\infty} h(\tau) e^{-j\omega \tau} d\tau = H(j\omega) se^{j\omega t}
\end{aligned}
$$

The above observation suggests that complex exponential signals can be used as waveforms for multipath channels. On the other hand, it is important to realize that the above relation is only valid for an infinitely long complex exponential. For practical digital communication, the signal waveforms must be confined to a symbol period.

Fortunately, if the channel response is FIR, i.e., $h(t) = 0, t \notin [\ 0 \ \ \tau_{\max}\ ]$, the same input-output property holds within a finite observation window duration T. The idea is illustrated in Figures 2.5 and 2.6. In Figure 2.5, a zero-padded guard period is inserted between neighboring symbols to prevent intersymbol

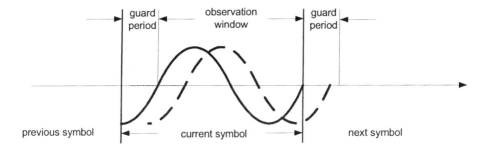

Figure 2.6: Cyclic-prefix guard interval

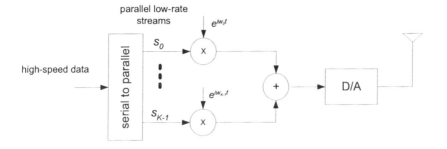

Figure 2.7: FDM scheme for high-speed transmission

interference (ISI). However the integrity of the waveform is not preserved[1]. Instead of using a silent period to guard against the ISI, a *cyclic-perfix (CP)* of duration at least $\tau_{\max}$ can be employed at the transmitter. As shown in Figure 2.6, any multipath components with delays less or equal to $\tau_{\max}$ will maintain their complex exponential waveforms within the observation window, leading to an intact signal waveform (within a scalar ambiguity) at the receiver side.

To carry more information in a given time window, the old-fashioned frequency-division multiplexing (FDM) can be utilized as shown in Figure 2.7. In particular, K information-bearing symbols, $s_0, \cdots, s_{K-1}$, can be modulated onto K different subchannels using different complex exponential $w_k(t) = e^{j2\pi kt/N}$ as follows.

$$x(t) = \sum_{k=0}^{K-1} s_k w_k(t), \quad t \in [\ -\tau_{\max}, \quad T\ ] .$$

---

[1]If the starting point of the observation window is perfectly known, it is possible to restore the waveform by patching the distorted portion with samples within the guard period of the next symbol.

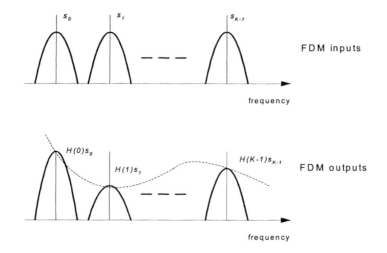

Figure 2.8: The input–output relation of FDM signals

With the CP, the output is only subject to a scalar multiplication on each of the information symbols within the observation window.

$$y(t) = \sum_{k=0}^{K-1} H(e^{j2\pi kt/N}) s_k w_k(t)$$

Graphically, the effect of the channel is a mere "scaling" on each subchannel as shown in Figure 2.8. Since the scalar ambiguity can be removed with channel estimation, it can be argued that the FDM is "immune" to the time-dispersion effect and thus has an advantage over single-carrier modulation with a linear equalizer.

Note however, such immunity is achieved at the expense of an unused CP (at the receiver). The ratio of $\tau_{\max}/T$ determines the signal overhead of the system. Since $\tau_{\max}$ is often fixed in a given application, there is an incentive to increase T in order to increase the system efficiency. The tradeoff though, is increased sensitivity to frequency offset - see Chapter 3.

Q2: How tight can we pack the signals in a given bandwidth?

While the CP in FDM reduces the multipath channel effect to a scalar on each subchannel, it is unclear whether the signals across different subchannels will interfere to each other. For two waveforms to be orthogonal within $[0, T]$, they must fulfill the orthogonality constraint:

$$\int_0^T w_1(t) w_2^*(t) dt = 0$$

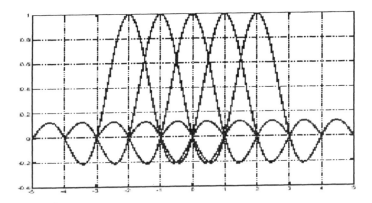

Figure 2.9: The spectra of OFDM signal

The minimum tone separation is thus $1/T$ hertz. Therefore, as long as $w_k(t)$ is placed $1/T$ hertz apart, there will be no inter-subchannel interference in FDM. The FDM that satisfies such frequency spacing requirement is therefore referred to as OFDM - orthogonal frequency division multiplexing. Each tone in OFDM is referred to as a *subcarrier*. The typical spectrum of OFDM signals is depicted in Figure 2.9. The overlapping *sinc* shaped spectra assure zero inter-subchannel interference at the right frequency sampling points.

Q3: *Does fast implementation exist for broadband modem?*

We now consider the digital implementation and the complexity issue of the OFDM modem.

Assume there are K (usually power of 2) subcarriers in the system, and the time-domain sampling rate is $N/T = 1/T_s$, $N = K$ hertz. Further assume that the channel delay spread is L ($LT_s = \tau_{\max}$) samples. Let x_n, h_n and y_n be the sampled input, channel, and output, respectively; it is clear that they are related by a linear convolution:

$$y_n = x_n * h_n$$

Now let $Y(k), X(k)$, and $H(k)$ be the K-point Discrete Fourier transforms (DFTs) of the output, input, and channel, respectively, within the observation window. In ordinary cases,

$$Y(k) \neq X(k)H(k), \qquad k = 0, ..., K - 1$$

since the y_n is related to x_n and h_n by a linear convolution not a *circular* convolution. However by appending an L point CP, $x_{N-L}, x_{N-L+1}, \cdots, x_{N-1}$, to x_n, the circular convolutional effect is created within the N point time window (please prove this as an exercise).

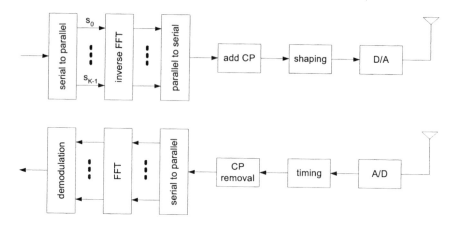

Figure 2.10: OFDM transceiver block diagram

In other words, in discrete frequency domain the channel effect is also a multiplicative constant:

$$Y(k) = X(k)H(k), \qquad k = 0, ..., K - 1 \qquad (2.5)$$

As a result, the OFDM modem can be elegently implemented in discrete-time using fast algorithms such as the Fast Fourier transform (FFT). Signals modulated on different subcarriers can be perfectly separated after the FFT operation at the receiver.

2.3 OFDM realization

Figure 2.10 depicts a typical transmitter and receiver chain of an OFDM modem. Unlike signal-carrier modulation, the OFDM modem is performed on a *block-by-block* basis. At the transmitter, a block of information-bearing symbols are first serial-to-parallel converted onto K *subcarriers*. The orthogonal waveform modulation is carried out using an inverse FFT and a parallel-to-serial converter. Following the converter, the last L points are appended to the beginning of the sequence as the cyclic prefix. The resulting samples are then shaped and transmitted. Each transmmited block is referred to as an *OFDM symbol*.

The receiver reverses the process using an FFT operation. In particular, the sampled signals are first processed to determine the starting point of a block and the proper demodulation window. By removing the CP (which now contains ISI), an N ($N = K$) point sequence is serial-to-parallel converted and fed to the FFT. The output of the FFT are the symbols modulated on the K subcarriers, each multiplied by a complex channel gain. Depending the availability of the

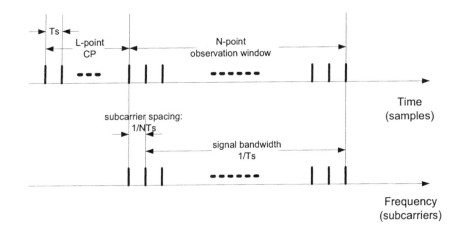

Figure 2.11: OFDM representation in time and frequency domains

channel information, different demodulation/decoding schemes are then used to recover the information bits.

Considering an OFDM system with N subcarriers and a time-domain sampling rate $1/T_s$, a time and frequency representation of an OFDM symbol is depicted in Figure 2.11.

Now let us examine the modem operations mathematically. Denote

$$\mathbf{s}(n) = [s_0(n) \ s_1(n) \cdots s_{P-1}(n)]^T$$

as the nth block of data to be transmitted. The number of subcarriers used, P, may be less or equal to the number of total available subcarriers, K: $P \leq K$. The OFDM modulation is implemented by applying an IDFT operator to the data stream $\mathbf{s}(n)$. Using matrix representation, the resulting N-point time domain signal is given by

$$\mathbf{x}(n) = [x_0(n) \ x_1(n) \cdots x_{N-1}(n)]^T = \mathbf{W}_P \mathbf{s}(n),$$

where $\mathbf{W}_P$ is an $N \times P$ submatrix of the IDFT matrix $\mathbf{W}$. The columns of $\mathbf{W}_P$ correspond to the subcarriers that are modulated with data.

For DFT-based OFDM, a cyclic prefix is added to the multiplexed output of the IDFT to form an OFDM symbol, before it is transmitted through a fading channel. Because of (2.5), the receiver output for the nth block within the demodulation window is given by

$$\begin{aligned}
\mathbf{y}(n) &= [y_0(n)\ y_1(n)\cdots y_{N-1}(n)]^T & (2.6)\\
&= \mathbf{W}_P\mathbf{H}\,\mathbf{s}(n) & (2.7)\\
&= \mathbf{W}_P\begin{bmatrix} H(1) & & 0 \\ & \ddots & \\ 0 & & H(P) \end{bmatrix}\mathbf{s}(n), & (2.8)
\end{aligned}$$

where $H(i)$, $i = 1\cdots,N$ is DFT of the channel response. In other words, each subchannel, with a scalar ambiguity, can be recovered by applying a DFT to $\mathbf{y}(k)$:

$$\mathbf{W}_P^H\mathbf{y}(n) = \mathbf{H}\,[s_1(n)\cdots s_P(n)]^T \qquad (2.9)$$

In practice, several additional operations are often needed at the transmitter and the receivers:

- *Cyclic prefix and postfix:* the cyclic prefix provides a guard interval for all multipath following the first arrival signal. As a result the timing requirement of the observation window is quite relaxed (up to $\tau_{\max}$ ambiguity). On the other hand, timing estimation often hinges on the multipath signal with the highest strength, which in some cases may not be the first arrival signal. To increase the robustness of the receiver, the guard interval is often split into cyclic pre-fix and post-fix as in Figure 2.12 to guard against early and late (relative to the strongest path) multipath signals.

- *Pulse shaping:* since the time-limited signal waveforms have strong raised-cosine sidelobes in the frequency-domain, OFDM has been shown to be sensitive to frequency offset which leads to inter-carrier interference. The rectangular time window also leads to high out-of-band emission (Figure 2.9) which is undesirable in radio communications. An effective way to reduce the ICI sensitivity is to pulse shape (time domain multiplication) the OFDM symbol with a pulse-shaping window. The tradeoff is a reduced guard-inteval and increased complexity. Alternatively, one can apply filters to limit the spectrum of the OFDM signals. Notice that filtering introduces the same convolutional effects as the multipath channel, therefore reduces system tolerance to delay spread given a CP.

- *Virtual carriers:* In order to guard against neightboring band interference and the out-of-band emission, a portion of the subcarriers at the two edges of the band may not be modulated. These unused subcarriers are termed as the *virtual carriers*. The concept is illustrated in Figure 2.13. As a result, the number of subchannels that carry the information is generally smaller than the size of the DFT block, *i.e.*, $P < N = K$. The virtual carriers provide a guard band for neighboring channels. In the absence of

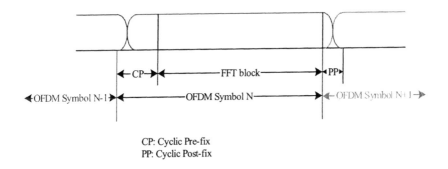

CP: Cyclic Pre-fix
PP: Cyclic Post-fix

Figure 2.12: OFDM symbols with cyclic pre-fix and post-fix

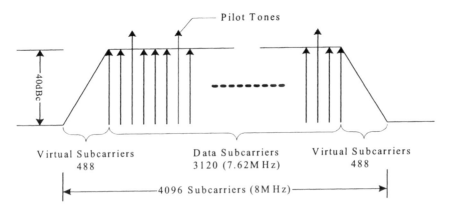

Figure 2.13: OFDM with virtual carriers

adjacent-channel-interference (ACI), the outputs from these virtual carriers are zero. For simplicity, we shall use the logic indices 1 to P to denote data subcarriers and $P + 1$ to N to denote virtual carriers.

Example 1 *OFDM parameter selection in IEEE 802.11a: RF measurements have shown that in a typical LAN environment, the rms delay spread at 5GHz is on the order of 50ns, with maximum delay up to 500ns.*

The cyclic guard interval in 802.11a is conservatively set at 800 ns (out of each 4μs OFDM symbol interval). From (2.3), the coherence bandwith is on the order of 4MHz. The subcarrier spacing in 802.11a is chosen to be 0.3125 Mhz, which is far less than the coherence bandwidth. This setting guarantees flat fading on individual subcarriers.

Out of a size 64-FFT, 12 are virtual carriers while the rest 52 subcarriers are dedicated to data (48) and pilots (4).

Example 2 *The DVB-T parameters: Depending on the region, the DVB-T bandwidth can be 6, 7, or 8Mhz. The length of the cyclic guard interval must be selected to accommodate "single-frequency-network" (SFN), where neighboring TV towers deliver the same OFDM modulated signals. The SNF creates artificial multipath with delay spread dictated by the maximum permissible distance between neighboring transmitters (e.g., 60 km). Accordingly, the DVB-T supports two modes of operations: the 2K-FFT and the 8K-FFT, with six possible values for guard interval from 224 us to 7 us. SNF with spacing up to 67 km can be realized with the longest CP (224 us ×300, 000 km/s ≈ 67 km).*

The 8K-FFT mode has subcarrier spacing 1.25 KHz, and a significantly longer symbol duration (e.g., 1 ms) than that in the 2K-FFT mode. In general, the 8K-FFT mode can only support fixed applications whereas the 2K-FFT mode is more suitable for portable and possibly mobile services.

2.4   Summary

The salient features of OFDM are highlighted as follows:

- Advantages

  - High spectral efficiency: OFDM is a highly efficienct modulation scheme which has been shown to approach the information threotical capacity with water-filling across its subcarriers. Although subcarrier-based power loading is less feasible in practice, adaptive coded modulation on OFDM subchannels (each subchannel comprises of a group of subcarriers) has already been adopted into IEEE standards.

  - Simple implementation: the use of FFT and IFFT in OFDM reduces the modem complexity, especially at the receiver. With the FFT, the number of operations in each OFDM symbol is on the order of $N \log N$. The implementation complexity of single carrier system with an equalizer is at least $N L_e$, where L_e is the number of taps in the equalizer.

  - Resistance to fading and interference: OFDM is robust against frequency selective fading and interference. With channel state information, maximum likelyhood detection can be effectively implemented for any given channel profile or interference pattern.

- Disadvantages:

  - Sensitivity to frequency offset: OFDM has strong tolerance against timing offset because of the cyclic guard interval. On the other hand, its tightly packed subcarriers give rise to increased sensitivity with respect to carrier frequency errors (oscillator impairments etc.) and *interchannel interference* (ICI). The extend of performance degradation due to carrier frequency offset is a function of the subcarrier spacing, the size of the FFT, and the modulation schemes used.

     – High peak-to-average power ratio: since the OFDM signal is the superposition of low rate streams modulated at different frequencies, its time-domain dynamic range increases with the number of subcarriers. The high peak-to-average power ratio (PAPR) imposes stringent requirements on the A/Ds and D/As, and more importantly, on the linearity of the power amplifier (PA).

In this chapter, we describe the challenges in broadband communications by characterizing the wireless fading channels. The statistical model of time- and frequency selective fading channels is presented. Based on a multi-carrier communication framework, we introduce the concept of OFDM and explain how channel fading can be mitigated with OFDM modulation. The canonical block diagram of an OFDM modem is described, along with implementation details, such as cyclic pre- and postfix, virtual carriers, and pulse shaping essential to a practical OFDM realization.

Bibliography

[1] P. Hoeher, "A statistical discrete-time model for the WSSUS multipath channel," *IEEE Trans. Communications*, 41(4): 461-468, November 1992.

[2] T. S. Rappaport, *Wireless Communications: Principles & Practice*, Prentice-Hall, Englewood Cliffs, NJ, 1996.

[3] J.A.C. Bingham, "Multicarrier Modulation for Data Transmission: An Idea Whose Time Has Come," *IEEE Communications Magazine*, 28, 5, pp. 5-14.

[4] L. J. Cimini, "Analysis and Simulation of a Digital Mobile Channel Using Orthogonal Frequency-Division Multiplexing," *IEEE Transactions on Communications*, 33, 7, pp. 665-675, 1985.

[5] B. Le Floch, M. Alard, and C. Berrou, "Coded Orthogonal Frequency Division Multiplex," *Proceedings of the IEEE*, 83, 6, pp. 982-996, 1995.

[6] S. Hara, and R. Prasad, "Overview of Multicarrier CDMA," *IEEE Communications Magazine*, 35, 12, pp. 126-133, 1997.

[7] S. B. Weinstein and P. M. Ebert, "Data transmission by frequency division multiplexing using the discrete fourier transform," *IEEE Trans. Communications*, vol. COM-19, pp. 628-634, Oct. 1971.

[8] B. Hirosaki, "An Orthogonally Multiplexed QAM System Using the Discrete Fourier Transform," *IEEE Trans. Communications*, 29, 7, pp. 982-989, 1981.

[9] W.Y. Zou, and Y. Wu, "COFDM: an Overview," *IEEE Trans. Broadcasting*, 41, 1, pp. 1-8, 1995.

Chapter 3

PHY Layer Issues — System Imperfections

In this chapter, we investigate signal processing solutions for a few performance-limiting system imperfections in OFDM communications. We first identify a set of key problems whose solutions will enable OFDM radio networks to deliver on its potential; the rest thus balances fundamental investigations with system design and implementation aspects.

To effectively modulate and demodulate OFDM signals, a number of pre- and post-modem tasks must be performed. Signal processing plays an essential role in these tasks. The most important ones among others include

- frequency synchronization

- channel estimation

- phase noise compensation

- peak-to-average power radio (PAPR) reduction

- I/Q imbalance compensation

3.1 Frequency synchronization

The inherent immunity of OFDM to multipath comes at the price of increased sensitivity to inaccurate frequency reference [4] - contrasting to its single-carrier counterparts which are robust to frequency offset but critical to timing inaccuracy. A carrier offset at the OFDM receiver can cause losses in subcarrier orthogonality, and thus introduces interchannel interference (ICI) and severely degrades the system performance [4]. High accuracy carrier offset estimation and compensation is of paramount importance in OFDM communications.

Similar to other communication systems, carrier synchronization in OFDM is usually carried out in two phases, namely, acquisition and tracking. While the acquisition range is the focus during the initial phase, accuracy and stability is the more important design criterion during the tracking stage. In addition, the computational requirements from these two modes are also different. While high cost algorithms are affordable during acquisition, more computationally efficient methods are necessary for the tracking mode.

In many traditional communication systems, a time-domain pilot sequence, $p(1), p(2), \cdots, p(N)$, is inserted periodically and the received signal in the presence of a carrier offset will be

$$p(1)e^{j\phi}, p(2)e^{j\phi+\Delta\omega}, \cdots, p(N)e^{j\phi+\Delta\omega(N-1)} \tag{3.1}$$

The estimation of $\Delta\omega$ is a classic harmonic estimate problem [28] and therefore will not be discussed here. The remainder of this section discusses carrier offset estimation methodologies that exploit the OFDM-specific structural information.

In general, there are two types of OFDM carrier compensation approaches at the baseband, namely, pilot-based and non-pilot based. To ensure an adequate acquisition range, many practical OFDM systems empoly concentrated pilot OFDM symbols (e.g., pilots in IEEE 802.11a and preambles in IEEE 802.16) for initial acquisition. In addition, continuous *pilot subcarriers* are also available for frequency tracking purposes. The obvious advantage of pilot-based approaches is its reliability and accuracy. The non-pilot (or blind) methods estimate the carrier offset by exploiting some known structural information of an OFDM sequence, therefore eliminating the pilot overhead of the systems. Various blind techniques have been proposed to date [7].

In the following we first present the data model of OFDM signals in the presence of carrier offsets. The formulation provides important insight to the impact of imperfect carrier recovery on the demodulation performance. Based on the model established, we present two simple and yet powerful carrier recovery approaches suitable for real world applications.

3.1.1   OFDM carrier offset data model

In the presence of a carrier offset, the time samples of an OFDM signal are modulated by a residual frequency $e^{j\phi n}$. From (2.6), the output vector within the nth demodulation window becomes

$$\mathbf{y}(n) = \mathbf{E}\mathbf{W}_P\mathbf{H}\,\mathbf{s}(n)e^{j\phi(N+L)(n-1)} \tag{3.2}$$

where L is the length of the cyclic prefix and

$$\mathbf{E} = \text{diag}(1, \ e^{j\phi}, \cdots, e^{j(N-1)\phi})$$

In the absence of a carrier offset, applying FFT to the received vector separates signals modulated on different subcarriers as seen in Equation (2.9). On the other hand, since

$$\mathbf{W}_P^H \mathbf{y}(n) = \underbrace{\mathbf{W}_P^H \mathbf{E} \mathbf{W}_P}_{\neq \mathbf{I}} \mathbf{H} \ \mathbf{s}(n) e^{j\phi(N+L)(n-1)}$$

the carrier offset $\mathbf{E}$ matrix destroys the orthogonality among the subchannels. The output on each subcarrier is contaminated by interchannel interference (ICI) from neighboring subcarriers. To recover $\{\mathbf{s}(n)\}$, the carrier offset, ϕ, needs to be estimated and compensated *before* performing the DFT.

3.1.2 Pilot-based estimation

The initial carrier acquisition in OFDM often relies on largely concentrated pilots that are periodically inserted. The simplest scenario involves one or more pilot OFDM symbol(s) with the signal vector $\mathbf{s}$ (modulated on all subcarriers) known to the receiver. In IEEE 802.11a/g and IEEE 802.16d/e for example, dedicated pilots and preambles are available for timing and frequency synchronization.

Assume one pilot OFDM symbol is available. From (3.2), the receiver signal vector is given by

$$
\begin{aligned}
\mathbf{y} &= \mathbf{E}(\phi)\mathbf{W}_P \mathbf{H} \ \mathbf{s} + \mathbf{n} \\
&= \mathbf{E}(\phi)\mathbf{W}_P \mathbf{S} \mathbf{h} + \mathbf{n} \\
&= \mathbf{U}(\phi)\mathbf{h} + \mathbf{n}
\end{aligned}
$$

Here we introduce

$$
\begin{aligned}
\mathbf{S} &= \text{diag}(s_1, \ s_2, \cdots, s_P) \\
\mathbf{U}(\phi) &= \mathbf{E}(\phi)\mathbf{W}_P \mathbf{S}
\end{aligned}
$$

to facilitate the discussion. $\mathbf{S}$ is known to the receiver whereas $\mathbf{h}$ is the unknown channel vector.

The Maximum Likelihood (ML) estimator for ϕ and $\mathbf{h}$ is

$$(\widehat{\phi}, \widehat{\mathbf{h}}) = \arg_{\phi, \mathbf{h}} \min \| \mathbf{y} - \mathbf{U}(\phi)\mathbf{h} \|^2$$

Given that $\mathbf{h}$ is linear with respect to the above minimization, we can express $\mathbf{h} = \mathbf{U}^\dagger(\phi)\mathbf{y}$ and plug it back into the ML estimator as

$$\widehat{\phi} = \arg_\phi \min \| \mathbf{y} - \mathbf{U}(\phi)\mathbf{U}^\dagger(\phi)\mathbf{y} \|^2$$

The above is a one-dimensional minimization which could be carried out by numerical search.

For tracking purposes, pilots are often only available on a subset of subcarriers locations. In this case, the estimator has to cope with both the unknown channel and the unknown ICI from neighboring subcarriers that carry data information. Joint estimation of all these unknowns is challenging. Fortunately the residual carrier offset during tracking is usually small. Therefore, the ICI can be regarded as additional Gaussian noise.

Under this assumption, (3.2) can be approximated as

$$\mathbf{y}(n) = \mathbf{W}_P \mathbf{H} \; \mathbf{s}(n) e^{j\phi(N+L)(n-1)} + \widetilde{\mathbf{n}}(n)$$

where $\widetilde{\mathbf{n}}(n)$ accounts for both the noise and contributions from the ICI. The first term includes pilot signals that are not affected by the ICI.

Let $\mathcal{P}$ denote the subcarrier indices corresponding to the pilot locations. Clearly, at the location of the pilot subcarriers

$$\mathbf{w}_j^H \mathbf{y}(n) = H(j) s_j(n) e^{j\phi(N+L)(n-1)}, \quad j \in \mathcal{P}.$$

Therefore, one can time differentiate consecutive pilots of the subcarriers in $\mathcal{P}$ to obtain

$$\frac{\mathbf{w}_j^H \mathbf{y}(n)/s_j(n)}{\mathbf{w}_j^H \mathbf{y}(n+1)/s_j(n+1)} = e^{-j\phi(N+L)}, j \in \mathcal{P}; n = 1, \dots$$

The residual carrier offset can be readily estimated.

In order to reduce the noise amplification effect due to deep fading channels, the carrier offset estimation must be weighted across all pilot subcarriers based on the channel conditions $H(j), j \in \mathcal{P}$. In addition, smoothing across time n may be unnecessary to avoid instability - see adaptive filtering for more information [11].

3.1.3   Non-pilot based estimation

Non-pilot based approaches can achieve carrier offset estimation without using reference symbols. Instead of using pilot symbols, the frequency offset can also be estimated blindly by exploiting some structural and statistical information of the received signals. For example, Schmidl and Cox proposed a frequency synchronization scheme using OFDM symbols with identical halves [2]. J. van de Beek et al developed an ML estimator by exploiting the redundancy in the cyclic prefix [3].

The technique described here provides a high accuracy carrier estimate by taking advantage of the inherent orthogonality among OFDM subchannels [5].

Indeed, even when the OFDM signal is distorted by an unknown carrier offset, the received signal possesses an algebraic structure which is a direct function of the carrier offset. In particular it is pointed out in Section 2.3 that virtual carriers are often used in OFDM modems to provide guard intervals against neighboring bands. In IEEE 802.11/a/g for example, only 52 out of the 64 subcarriers are used for data modulation. The rest 12 are virtual carriers. In IEEE 802.16e, the number of virtual carriers exceeds 12% of the total number of subcarriers. We show in the following that this property permits the formulation of a cost function which yields a closed-form estimate of the carrier offset.

Recall from (2.9) that $\mathbf{W}_P$ consists of columns of the IDFT matrix corresponding to the used subcarriers. Its orthogonal complement,

$$\mathbf{W}_P^{\perp} = [\mathbf{w}_{P+1} \cdots \mathbf{w}_N],$$

contains columns of the IDFT matrix corresponding to the virtual carriers. Hence, in the absence of the carrier offset, $i.e.$, $\phi = 0$,

$$\mathbf{w}_{P+i}^H \mathbf{y}(n) = \mathbf{w}_{P+i}^H \mathbf{W}_P \mathbf{H}\,\mathbf{s}(n) = 0, \quad i = 1, \cdots, N - P.$$

Such is not true when $\phi \neq 0$. However, if we let

$$\mathbf{Z} = \begin{bmatrix} 1 & & & 0 \\ & z & & \\ & & \ddots & \\ 0 & & & z^{N-1} \end{bmatrix},$$

it can be easily shown that when $\mathbf{Z} = \mathbf{E}$,

$$\mathbf{w}_{P+i}^H \mathbf{Z}^{-1} \mathbf{y}(n) = \mathbf{w}_{P+i}^H \mathbf{Z}^{-1} \mathbf{E} \mathbf{W}_P \mathbf{H}\,\mathbf{s}(n) = 0 \quad i = 1, \cdots, N - P.$$

This observation suggests that we form a cost function given a finite number of data vectors as follows:

$$P(z) \;=\; \sum_{i=1}^{K-P} \sum_{n=1}^{N_p} \left\| \mathbf{w}_{P+i}^H \mathbf{Z}^{-1} \mathbf{y}(n) \right\|^2 \tag{3.3}$$

$$=\; \sum_{i=1}^{K-P} \sum_{n=1}^{N_p} \mathbf{w}_{P+i}^H \mathbf{Z}^{-1} \mathbf{y}(n) \mathbf{y}^H(n) \mathbf{Z} \mathbf{w}_{P+i} \tag{3.4}$$

In a system with many virtual carriers, we may choose a portion of the $N - P$ virtual carriers to reduce computational complexity without loss of performance. Clearly, $P(z)$ is zero when $z = e^{j\phi}$. Therefore one can find the carrier offset by evaluating $P(\phi)$ along the unit circle, as in the well-known MUSIC algorithm in array signal processing [6]. On the other hand, it is noted that $P(z)$ forms a polynomial of z with order $2(N - 1)$. This allows a closed-form estimate of ϕ through polynomial rooting. In particular, $e^{j\phi}$ can be identified as the root of $P(z)$ on the unit circle.

The algorithm is summarized as follows:

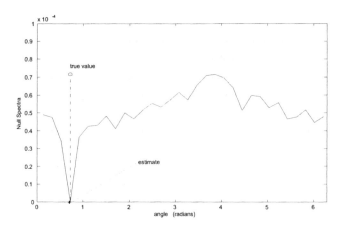

Figure 3.1: The null spectrum of carrier offset estimator

1. Form the polynomial cost function as in (3.3) using the receiver outputs, $\{\mathbf{y}(n)\}$.

2. Estimate the carrier offset as the null of $P(z)$ or the phase of the root of $P(z)$ on the unit circle. In the presence of noise, the carrier offset is estimated as the minima of $P(\phi)$ or the phase of the root of $P(z)$ closest to the unit circle.

The above algorithm takes advantage of the *known* structure of the subspace of OFDM signals and thus offers performance comparable to the subspace-based algorithms with minimum cost. It has been shown that the method is indeed the maximum likelihood (ML) estimate of the carrier frequency offset with a virtual carrier present signal model [8]. More recently, nonlinear least-squares estimators that are robust to correlated noise have been proposed [10].

Example 3 *An example OFDM system with $N = 32$ carriers and $P = 20$ data streams is considered. The carrier offset is estimated using virtual carrier based algorithm described in the previous section. The true frequency offset $\phi = 3.67\Delta\omega$, where $\Delta\omega = \frac{2\pi}{N}$ is the channel spacing. The estimation results using 4 symbol blocks of noise-free data are shown in Fig. 3.1 and Fig. 3.2 In particular Fig. 3.1 shows the null spectrum and Fig. 3.2 displays the root distribution of $P(z)$.*

The minimization of $P(\phi)$ in (3.3) for carrier offset estimation can be implemented adaptively, which readily leads to a carrier tracking algorithm for time-varying environments. Note that no matrix inversion is involved in the cost function, thus excellent transient behaviors of recusive least-squares (RLS) type of adaptive algorithms can be expected with low complexity.

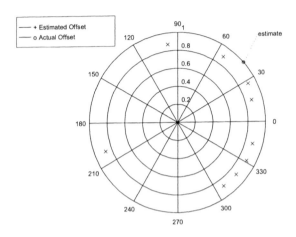

Figure 3.2: The root distribution of polynomial carrier offset estimator

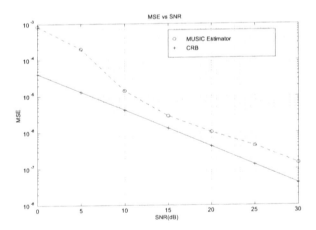

Figure 3.3: Performance of virtual carrier based carrier offset estimator

Example 4 *Figure 3.3 shows the performance of the virtual carrier based frequency estimator. The algorithm uses 1 block of data of size $N = 128$, $P = 68$ and $L = 5$. The true frequency offset $\phi = \Delta\omega$ equals the channel spacing. The mean-squared error (MSE) of the carrier offset estimate was obtained by running 200 independent realizations. As seen, the algorithm provides, under reasonable SNR values, very good performance.*

3.2 Channel estimation

To coherently demodulate OFDM signals, channel state information (CSI) must be available at the receiver side. For mobile applications where the radio channel is both time- and frequency-selective (see Section 2.1.4), dynamic channel estimation is necessary. The literature on OFDM channel estimation is abundant, most of which addresses channel estimation based on pilot tones inserted in the OFDM symbol stream. In [12] for example, the optimum pilot tone spacing is investigated assuming a first order Markov channel. Least-squares (LS) or minimum mean-square error (MMSE) based estimators and their low cost implementation are studied in [13],[14],[17]. Space-time channel estimation suitable for space-time fading channels is investigated in [15]. Parametric-based approaches, which further exploit the channel structure information for better estimation performance, can be found in [16] and reference therein.

In this section, we discuss the basics of statistical OFDM channel estimation and its implementation. Starting with the time-varying OFDM channel model, we first establish the minimum set of conditions for scattered pilots with which channel estimation can be performed. The optimum MMSE channel estimation is then described [13][14], followed by reduced complexity channel estimators that are more suitable for practical implementation.

3.2.1 Pilots for 2D OFDM channel estimation

In reference to Figure 2.11, each OFDM symbol has duration

$$T_{symbol} = (N + L)T_s \text{ seconds}$$

and the subcarrier spacing

$$f_{subcarrier} = 1/NT_s \text{ Hertz}$$

To facilitate channel estimation, scattered pilots are inserted in the time-frequency grid at pre-determined intervals. An example pilot pattern is illustrated in Figure 3.4. The time (in terms of OFDM symbols) and frequency (in terms of subcarrier spacing) intervals are largely determined by the characteristics of the channels, and in particular its maximum Doppler frequency and the multipath delay spread.

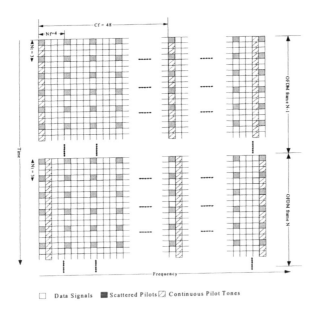

Figure 3.4: An example OFDM pilot structure

Let us assume that

1. The maximum frequency shift is F_D Hz, half of the Doppler bandwidth B_D;

2. The maximum delay spread $\tau_{\max} = LT_s$ seconds;

3. Discrete pilot symbols are placed N_f subcarriers apart in frequency and N_t OFDM symbols apart in time.

To fulfill the sampling theorem in both the time and frequency domains, the pilot spacing (N_f, N_t) must satisfy

$$\text{frequency spacing: } N_f \times f_{subcarrier} < \frac{1}{\tau_{\max}} \Rightarrow N_f < \frac{1}{\tau_{\max} \times f_{subcarrier}} \quad (3.5)$$

$$\text{time spacing: } \frac{1}{N_t T_{symbol}} > 2F_D \Rightarrow N_t < \frac{1}{2T_{symbol} \times F_D} \quad (3.6)$$

Example 5 *The IEEE 802.16e is intended to provide fixed, portable, and mobile broadband wireless services in a wide-area environment with delay spread up* *to $\tau_{max} = 20\ \mu s$. The maximum mobility speed anticipated is $v = 125 km/hr$. In the orthogonal frequency division multiple-access (OFDMA) mode, the subcarrier spacing is designed to be $f_{subcarrier} = 11.16\ KHz$ while the OFDM symbol duration is $100\ \mu s$[7].*

- *The maximum frequency domain spacing for the pilots is*

$$N_f < \frac{1}{\tau_{max} \times f_{subcarrier}} = 5\ [subcarriers]$$

- *The Doppler shift at 3.5 GHz carrier frequency is*

$$F_D = \frac{v}{\lambda} = \frac{35m/s}{0.086m} = 408Hz.$$

Therefore, the maximum time domain spacing for the pilots is

$$N_t < \frac{1}{2T_{symbol} \times F_D} = 5\ [OFDM\ symbols]$$

3.2.2   2D MMSE channel estimation

Assuming sufficient pilot symbols are inserted to satisfy the sampling theorem, the next step is to derive a statistical channel estimator for channel responses at all locations. In particular, we are interested in the linear minimum mean-squared error (LMMSE) channel estimator to best fit the channel statistics (known). Recall from (2.2) that the time-frequency channel response of the time-variant channel $h(f, t)$ can be modeled as random processes with covariance matrix:

$$R_h(\Delta f, \Delta t) = E\{h(f; t)h^*(f - \Delta f; t - \Delta t)\} \qquad (3.7)$$

The channel statistics can be obtained offline from channel modeling or online from data observations.

In discrete time and frequency, let k and l denote the subcarrier and OFDM symbol indices, respectively. The time-frequency input-output relation in an OFDM system is given by

$$y(k, l) = x(k, l)h(k, l) + n(k, l) \qquad (3.8)$$

where $x(k, l), y(k, l)$ and $n(k, l)$ are the transmitted signal, the received signal, and the noise.

Denote the location set of the pilots as $\mathcal{P}$ with corresponding frequency (subcarrier) index m and time (OFDM symbol) index n : $(m, n) \in \mathcal{P}$. In most cases, $\mathcal{P}$ is a small subset of all possible $\{(k, l)\}$. The amount of the pilots presents a tradeoff between channel estimation accuracy and the bandwidth overhead.

Let $r(m, n) = y(m, n)/x(m, n)$ be the normalized received signals at the pilot locations:

$$r(m, n) = h(m, n) + v(m, n); (m, n) \in \mathcal{P} \tag{3.9}$$

The goal of channel estimation is to determine the entire channel response $h(k, l)$ based on a *linear combination* (using 2-D filtering) of the observed pilots:

$$\widehat{h}(k, l) = \sum_{(m,n) \in \mathcal{P}} w(k, l, m, n) r(m, n)$$

Here, $\{w(k, l, m, n)\}$ represent time-frequency variant 2-dimentional filter coefficients. $w(k, l, m, n)$ can be determined using the minimum mean-squared error (Wiener filter) criteria:

$$w(k, l, m, n) = \arg_w \min E\left[|h(k, l) - \widehat{h}(k, l)|^2\right]$$

Theorem 1 *The orthogonality principle: The optimum linear mean-squares estimate $Ax + B$ of random variable y is such that the estimation error $y - (Ax + B)$ is orthogonal to the observation x :*

$$E\{[y - (Ax + B)]x\} = 0.$$

For OFDM channel estimation, the optimal filter coefficients that minimize the MSE are obtained when

$$E\left[(h(k, l) - \widehat{h}(k, l))r^*(m, n)\right] = 0$$

Taking expectations:

$$\underbrace{E[h(k, l)r^*(m, n)]}_{\text{cross}} = \sum_{k', l' \in P} w(k, l, k', l') \underbrace{E[r(k', l')r^*(m, n)]}_{\text{auto}}$$

Both the cross-covariance and the auto-covariance matrices are known *a priori* from Equations (2.2) and (3.9). The coefficients $\{w(k, l, m, n)\}$ can thus be obtained by solving the above linear equation set.

A few observations regarding the 2 dimensional (2D) channel estimator are listed below

- High performance: The above 2D channel estimator is optimum in terms of the MSE given the channel statistics. While further improvements are possible (e.g., parametric based channel estimation where certain channel parameters are assumed to be known [16]), the tradeoff between performance and robustness needs to be carefully evaluated [17].

- High complexity: Note that the computation of the channel response $h(k, l)$ at each location involves 2D filtering of all pilot signals. In addition, the filter coefficients are time- and frequency-dependent as well. Although these coefficients may be calculated offline, the amount of storage and computations required may be prohibitive in practice.

3.2.3   Reduced complexity channel estimation

To alleviate the computational problem, one can either employ a low-rank 2-D estimator or use separable filters that are both 1-dimensional.

The low-rank 2D estimator essentially reduces the 2D filtering to the *vicinity* of the channel response [13]. More specifically,

$$\widehat{h}(k, l) = \sum_{(m,n) \in \mathcal{P}} w(k, l, m, n) r(m, n);$$

$$w(k, l, m, n) = 0; \text{ if } |k - m| > \Delta m \text{ or } |l - n| > \Delta n$$

Δm and Δn are selected based on the channel coherence time and coherence bandwidth, as well as the computation complexity the channel estimator can afford.

Alternatively, the channel estimation complexity can be reduced by partitioning the 2D filtering into separate time-domain filtering and frequency domain filtering. The order with which the two 1-D filtering operations are performed depends on the actual pilot pattern.

Assuming that time-domain filtering is to be carried out first. Dropping the frequency index (k and m) in (3.9) yields

$$r_t(n) = h_t(n) + v_t(n); n \in \mathcal{P}_t \qquad (3.10)$$

where $\mathcal{P}_t$ is the time-domain pilot location set. Define

$$\mathbf{h}_t = [\ h(1) \quad h(2) \quad \cdots \quad h(N)\ ]^T$$

$$\mathbf{r}_{\mathcal{P}_t} = [\ r(1) \quad r(2) \quad \cdots \quad r(|\mathcal{P}_t|)\ ]^T$$

$$= \mathbf{h}_{\mathcal{P}_t} + \mathbf{v}_{\mathcal{P}_t}$$

as channel responses at all locations and at noisy observations at pilot locations $\mathcal{P}_t$ ($\mathcal{P}_t \ll N$) respectively. The time domain MMSE channel estimation (Wiener filtering) is given by

$$\widehat{\mathbf{h}}_{t,MMSE} = \mathbf{R}_{\mathbf{h}_t \mathbf{h}_{\mathcal{P}_t}} \left(\mathbf{R}_{\mathbf{h}_{\mathcal{P}_t}} + \frac{\mathbf{I}}{SNR} \right)^{-1} \mathbf{r}_{\mathcal{P}_t} = \mathbf{F}_t \mathbf{r}_{\mathcal{P}_t} \qquad (3.11)$$

where $\mathbf{R}_{\mathbf{h}_t \mathbf{h}_{\mathcal{P}_t}}$ and $\mathbf{R}_{\mathbf{h}_{\mathcal{P}_t}}$ are the cross and auto channel correlation matrices

$$\mathbf{R}_{\mathbf{h}_t \mathbf{h}_{\mathcal{P}_t}} = E\{\mathbf{h}_t \mathbf{h}_{\mathcal{P}_t}^H\}$$
$$\mathbf{R}_{\mathbf{h}_{\mathcal{P}_t}} = E\{\mathbf{h}_{\mathcal{P}_t} \mathbf{h}_{\mathcal{P}_t}^H\}$$

$\mathbf{R}_{\mathbf{h}_t \mathbf{h}_{\mathcal{P}_t}}$ and $\mathbf{R}_{\mathbf{h}_{\mathcal{P}_t}}$ are determined by the channel Doppler spectrum and are assumed known to the receiver. Otherwise, the "robust" channel correlation should be used as suggested in [17] if no prior channel knowledge is available).

After time-domain channel estimation, the resulting estimates $\widehat{\mathbf{h}}_{t,MMSE}$ can be used as part of the pilot observations for frequency-domain channel estimation. The effective size of the frequency domain pilot set, $K = |\mathcal{P}_f|$, is therefore increased. In many cases, K is power-of-2 with a comb-type pilot pattern.

Following a similar derivation, we obtain the MMSE estimate of the frequency domain channel response vector as

$$\widehat{\mathbf{h}}_{f,MMSE} = \mathbf{R}_{\mathbf{h}_f \mathbf{h}_{\mathcal{P}_f}} \left(\mathbf{R}_{\mathbf{h}_{\mathcal{P}_f}} + \frac{\mathbf{I}}{SNR} \right)^{-1} \mathbf{r}_{\mathcal{P}_f} = \mathbf{F}_f \mathbf{r}_{\mathcal{P}_f}. \quad (3.12)$$

The computation of the above estimate may be prohibitive in practice if the number of subcarriers, N, is large. To reduce the computations, rearrange the frequency channel response vector into

$$\mathbf{h}_f = \begin{bmatrix} \mathbf{h}_{\overline{\mathcal{P}}_f} \\ \mathbf{h}_{\mathcal{P}_f} \end{bmatrix}$$

where $\mathbf{h}_{\mathcal{P}_f}$ corresponds at the pilot locations and $\mathbf{h}_{\overline{\mathcal{P}}_f}$ is the response vector at other locations. Clearly,

$$\widehat{\mathbf{h}}_{\mathcal{P}_f,MMSE} = \mathbf{R}_{\mathbf{h}_{\mathcal{P}_f}} \left(\mathbf{R}_{\mathbf{h}_{\mathcal{P}_f}} + \frac{\mathbf{I}}{SNR} \right)^{-1} \mathbf{r}_{\mathcal{P}_f} \quad (3.13)$$

$$\widehat{\mathbf{h}}_{\overline{\mathcal{P}}_f,MMSE} = \underbrace{\mathbf{R}_{\mathbf{h}_{\overline{\mathcal{P}}_f} \mathbf{h}_{\mathcal{P}_f}} \mathbf{R}_{\mathbf{h}_{\mathcal{P}_f}}^{-1}}_{\text{interpolation filter}} \widehat{\mathbf{h}}_{\mathcal{P}_f,MMSE}$$

In other words, the frequency channel responses at the pilot locations can be estimated first, whereas the rest can be obtained by interpolation based on these estimates.

Since $\mathbf{h}_{\mathcal{P}_f}$ is the Fourier transform of the finite duration channel impulse response (with most of its energy limited to the maximum delay spread), $\mathbf{R}_{\mathbf{h}_{\mathcal{P}_f}}$ is rank deficient in general. Performing an EVD on $\mathbf{R}_{\mathbf{h}_{\mathcal{P}_f}}$ yields

$$\mathbf{R}_{\mathbf{h}_{\mathcal{P}_f}} = \mathbf{U} \mathbf{\Lambda} \mathbf{U}^H$$

where $\mathbf{U}$ is a unitary matrix and $\mathbf{\Lambda}$ is a diagonal matrix containing $q < K$ non-zero eigen values $\{\lambda_i\}$. If we further assume that all multipath signals are independent, $\mathbf{U}$ reduces to a K-point FFT matrix $\mathbf{W}_K$. As a result,

$$\widehat{\mathbf{h}}_{\mathcal{P}_f, MMSE} = \mathbf{W}_K \begin{bmatrix} \delta_1 & & 0 \\ & \ddots & \\ 0 & & \delta_K \end{bmatrix} \mathbf{W}_K^H \mathbf{r}_{\mathcal{P}_f}$$

$$\delta_i = \begin{cases} \frac{\lambda_i}{\lambda_i + 1/SNR} & i = 1, \cdots, q \\ 0 & i = q+1, \cdots, K \end{cases}$$

Essentially the MMSE estimation is carried out in three steps

1. Step 1: converting the frequency domain observations to the time domain with IFFT: $\mathbf{W}_K^H \mathbf{r}_{\mathcal{P}_f}$;

2. Step 2: weighting and truncating the time-domain channel response based on the multipath delay profile;

3. Step 3: converting the resulting time-domain channel response back to frequency domain with FFT $\mathbf{W}_K$.

Once $\widehat{\mathbf{h}}_{\mathcal{P}_f, MMSE}$ becomes available, several different approaches can be employed to obtain $\widehat{\mathbf{h}}_{\overline{\mathcal{P}}_f}$ through interpolation:

- *MMSE based:* (3.13) provides the MMSE based interpolation that is computationally expensive.

- *FFT based:* The $|\mathcal{P}_f|$-point channel impulse response in Step 2 can be zero-padded into an N-point vector. The upsampled frequency channel response is calculated by applying FFT to the N-point time response vector.

- *Filter based:* linear low-pass interpolation filters can be applied directly to $\widehat{\mathbf{h}}_{\mathcal{P}_f, MMSE}$. Essentially the interpolation is performed by inserting $N - |\mathcal{P}_f|$ zeros into $\widehat{\mathbf{h}}_{\mathcal{P}_f, MMSE}$ and then applying a low pass FIR filter. The bandwidth of the filter should be selected based on the coherent bandwidth of the channel.

Figure 3.5 shows the block diagram.

Example 6 *In this example a wireless channel with Doppler shift $F_D = 200Hz$*

(120km/s @ 1.8G) and delay spread $Ts = 73$ us $(F_D * Ts = 0.015)$ is considered. An OFDM system with 256 subcarriers is used. The MSE performance of the channel estimation using three different channel estimation schemes is compared in Figure 3.6. As seen, the 2D algorithm utilizing prior knowledge of the channel yields the best performance. Its performance degrades below that of the 1D approach when no knowledge of the channel is available.

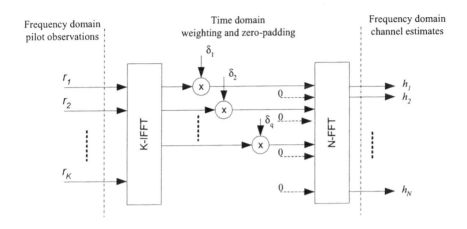

Figure 3.5: A block diagram of FFT-based channel estimation

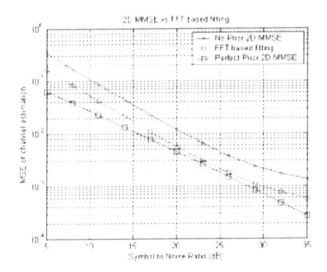

Figure 3.6: Channel estimation performance comparison

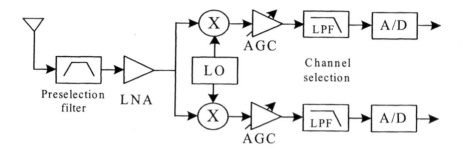

Figure 3.7: A direct-conversion receiver architecture

3.3   I/Q imbalance compensation

To maximize the spectrum efficiency, high-order modulation schemes (e.g., 64-QAM) are used in broadband wireless systems. The combination of OFDM and high order modulation imposes stringent requirements on RF devices [23]. This is particularly true for low-cost direct-conversion RF receivers that suffer from a higher degree of RF imperfections such as the DC-offset and the I/Q imbalance. Sophisticated signal processing algorithms are needed to cope with these RF imperfections in OFDM/high-order QAM.

This section addresses the I/Q imbalance problem in presence of frequency offset. The I/Q imbalance is commonly seen in any RF front-end that exploits analog quadrature down-mixing. This imbalance mainly attributes to the mismatched components in the I and the Q branches. Examples include, but are not limited to, an imperfectly balanced local oscillator (LO) and/or baseband low pass filters (LPF) with mismatched frequency responses. In contrast to an ideal down-converter that performs simple frequency shifting, a down-converter with I/Q imbalance not only down-converts the desired signal, but also introduces its image interference [19][24]. Such image interference, if left uncorrected, presents an error floor which limits the demodulation performance. Moreover, although the I/Q imbalance introduced by the LO may be assumed constant over the signal bandwidth, the mismatches in the subsequent baseband I/Q amplifiers and filters tend to vary with frequencies. Such frequency dependent I/Q imbalance is particularly severe in a wideband direct-conversion receiver and the corresponding estimation and compensation process becomes more challenging [20]. While abundant literature exists on I/Q imbalance compensation (see [19]-[25] and references therein), only a few of them consider the frequency dependent I/Q imbalance [24][20][22][21], let alone the frequency dependent I/Q imbalance in the presence of a frequency carrier offset.

In the following, we first establish a data model for frequency dependent I/Q imbalance and frequency offset. The discussion leads naturally to a receiver

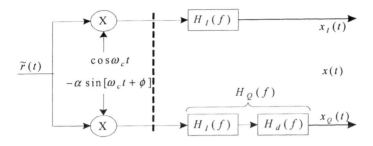

Figure 3.8: An receiver with I/Q imbalance

structure that can compensate both imperfections. To determine the compensation parameters, we introduce a pilot-based scheme for both frequency offset and I/Q imbalance estimation.

3.3.1   I/Q imbalance model

Figure 3.7 shows the general architecture of a direct-conversion receiver. Its mathematical model is given in Figure 3.8. We categorize the I/Q imbalance into a frequency independent part and a frequency dependent part. First, the imbalance caused by the LO can be characterized by an amplitude mismatch α and a phase error ϕ [19]. Since the LO generates a single tone, it is reasonable to model α and ϕ as frequency independent [24][22]. Following the LO are mixers, amplifiers, LPFs and A/D converters, which in general cause the frequency dependent I/Q imbalance. We represent the I/Q baseband signal paths by two mismatched LPFs (with frequency responses of $H_I(f)$ and $H_Q(f)$ respectively) [24]. Using the I channel frequency response as reference, the frequency dependent I/Q imbalance can be modeled by a difference term $H_d(f) = H_Q(f)/H_I(f)$. Since the amplitude mismatch α can be treated as part of the LPF difference $H_d(f)$, we will ignore the effect of α in the ensuing discussion and only consider the frequency independent phase error ϕ and the frequency dependent imbalance $H_d(f)$.

To understand the impact of I/Q imbalance on signal reception, we define the received signal as

$$\tilde{r}(t) = Re\{r(t) \times e^{j(\omega_c + \Delta\omega)t}\} \qquad (3.14)$$
$$r(t) = r_I(t) + j \times r_Q(t) = s(t) * c(t) \qquad (3.15)$$

where $r(t)$, $s(t)$ and $c(t)$ are the baseband representations of the received signal, the transmitted signal, and the channel response, respectively. $\tilde{r}(t)$ is the received RF signal modulated at center frequency ω_c with a frequency offset $\Delta\omega$.

Following the derivation in [24] and taking into account the initial frequency offset, the down-converted baseband signal $x(t)$ can be represented as

$$
\begin{aligned}
x(t) &= \mathcal{LPF}\{\tilde{r}(t)e^{-j\omega_c t}\} \qquad\qquad\qquad\qquad (3.16)\\
&= r(t)e^{j\Delta\omega t} * h_I(t) \otimes g_+(t) + r^*(t)e^{-j\Delta\omega t} * h_I(t) \otimes g_-(t)\\
&= e^{j\Delta\omega t}s(t) * d(t) + e^{-j\Delta\omega t}s^*(t) * v(t) \qquad\qquad (3.17)
\end{aligned}
$$

where

$$
\begin{aligned}
g_+(t) &= \mathcal{F}^{-1}\{G_+(f)\} = \mathcal{F}^{-1}\{[1 + e^{-j\phi} \cdot H_d(f)]/4\}\\
g_-(t) &= \mathcal{F}^{-1}\{G_-(f)\} = \mathcal{F}^{-1}\{[1 - e^{j\phi} \cdot H_d(f)]/4\}\\
h_I(t) &= \mathcal{F}^{-1}\{H_I(f)\}, \quad h_Q(t) = \mathcal{F}^{-1}\{H_Q(f)\}\\
d(t) &= c(t) * h_I(t) * g_+(t)\\
v(t) &= c^*(t) * h_I(t) * g_-(t).
\end{aligned} \qquad (3.18)
$$

Notice that different from the output of an ideal down-conversion, the desired signal is contaminated by its image interference as seen in Figure 3.9. $d(t)$ is the composite channel for the signal of interest and $v(t)$ quantifies the severity of I/Q imbalance. The signal to image interference ratio is given by

$$
SIR = \int_{-\infty}^{+\infty} |d(t)|^2 dt \Big/ \int_{-\infty}^{+\infty} |v(t)|^2 dt . \qquad (3.19)
$$

In practice, the typical image rejection is only around 30dB even with careful design.

After the down-converted signal is digitized at a rate that satisfies the Nyquist sampling theorem, the resulting discrete-time representation of (3.16) is given by

$$
x(n) = e^{j\Delta\omega T_s n}s(n) * d(n) + e^{-j\Delta\omega T_s n}s^*(n) * v(n) \qquad (3.20)
$$

with T_s being the sampling period.

3.3.2 Digital compensation receiver

The above model suggests a compensation receiver depicted in Figure 3.10 where the frequency independent imbalance, the frequency dependent imbalance, and the carrier offset are compensated individually.

Since the frequency dependent imbalance is attributed to the different LPF responses of analog I/Q branches, it can be balanced out with an L-order FIR filter $W(f)$. After the frequency dependent imbalance has been removed, the remaining frequency independent I/Q imbalance and the carrier offset effect can be characterized by a matrix Φ as follow:

$$
\begin{aligned}
\begin{bmatrix} x_I(n) \\ x_Q(n) \end{bmatrix} &= \begin{bmatrix} 1 & 0 \\ -\sin\phi & \cos\phi \end{bmatrix} \begin{bmatrix} \cos\Delta\omega T_s n & \sin\Delta\omega T_s n \\ -\sin\Delta\omega T_s n & \cos\Delta\omega T_s n \end{bmatrix} \begin{bmatrix} r_I(n) \\ r_Q(n) \end{bmatrix}\\
&= \Phi \begin{bmatrix} r_I(n) \\ r_Q(n) \end{bmatrix}
\end{aligned} \qquad (3.21)
$$

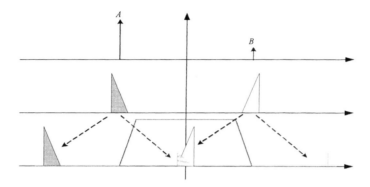

Figure 3.9: Interference caused by I/Q imbalance

where $[r_I(n) \; r_Q(n)]^T$ and $[x_I(n) \; x_Q(n)]^T$ are the baseband representations of the mixer input and output, respectively. The compensation is thus simply left-multiplying the received vector $[x_I(n) \; x_Q(n)]^T$ by $\boldsymbol{\Phi}^{-1}$, which can be factored into

$$
\boldsymbol{\Phi}^{-1} = \underbrace{\begin{bmatrix} \cos \Delta\omega T_s n & -\sin \Delta\omega T_s n \\ \sin \Delta\omega T_s n & \cos \Delta\omega T_s n \end{bmatrix}}_{frequency\ compensation} \underbrace{\begin{bmatrix} \cos\phi & -\sin\phi \\ \sin\phi & \cos\phi \end{bmatrix}}_{phase\ rotation} \begin{bmatrix} 1/\cos\phi & \tan\phi \\ 0 & 1 \end{bmatrix}.
$$

(3.22)

Therefore, the $\boldsymbol{\Phi}^{-1}$ compensation is composed of three terms: the left term corresponds to the frequency offset compensation; the middle term is just a phase rotation and will be absorbed in the channel equalization later on; and the right term suggests a compensation structure that corrects the frequency independent I/Q imbalance with only two multiplication and one addition. We denote this term of the I/Q compensation *asymmetric phase compensator* – similar to the asymmetric form in [25]. It should also be noted that the factorization in (3.22) indicates that the I/Q imbalance be compensated before the frequency offset compensation.

In Figure 3.10, the FIR filter $w(n)$ in the I branch is intended for compensating the frequency dependent imbalance. The $\delta(n - L/2)$ block in the Q branch is simply a delay unit for matching the delay introduced by $w(n)$. Between the dashed lines is the asymmetric phase compensator, where the gain factors a and b correspond to $1/\cos\phi$ and $\tan\phi$ respectively as suggested in (3.22). To further simplify the compensation structure, the gain factor a in the I branch can be merged into $w(n)$ as an additional scaling factor. It is also worth pointing out that without the frequency dependent I/Q imbalance, $w(n)$ reduces to a scalar and the corresponding I/Q compensation structure reduces to the asymmetric compensator as described in [25].

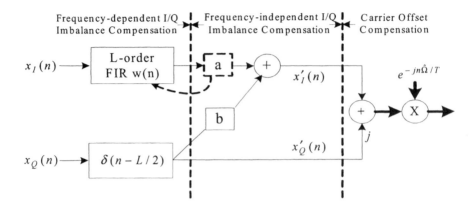

Figure 3.10: Receiver structure for I/Q and carrier offset compensation

3.3.3   Frequency offset estimation with I/Q imbalance

Due to the tangled effect of I/Q imbalance and the frequency offset, determining the parameters for the compensation receiver is nontrivial even with known training symbols. For frequency estimation, the synchronization pilot usually contains several identical symbols. Numerous estimators have been proposed accordingly, e.g., [5] and [26]. Despite their good performance, the accuracy of these estimators degrades in the presence of I/Q imbalance as will be shown later in this section. In the following, we introduce a pilot structure similar to the short SYNCs of IEEE 802.11a and reformulate the problem to take into account the I/Q imbalance. The reformulation leads to a nonlinear least-squares (NLS) based method that can provide accurate estimation even with severe I/Q imbalance.

Figure 3.11 shows the pilot structure for frequency offset and I/Q imbalance compensation. In particular, the pilot contains M identical symbols (each containing N samples, denoted as $p(n)$) with all the even symbols rotated by $\pi/4$. Since the successive pilot symbols are no longer strictly identical (due to the $\pi/4$ rotation), a guard interval (GI) or cyclic prefix (CP) of length P longer than the channel delay spread is introduced to avoid inter-symbol interference (ISI).

To appreciate this pilot pattern, we first examine M identical pilot symbols without the $\pi/4$ phase rotation between pilot symbols. The reason for the additional rotation will become obvious later.

After the GI/CP removal, we stack the received pilot samples in a matrix

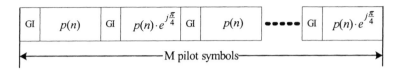

Figure 3.11: Pilot pattern for I/Q imbalance and frequency offset estimation

as follows:

$$X = \begin{bmatrix} x(1,1) & x(1,2) & \cdots & x(1,N) \\ x(2,1) & x(2,2) & \cdots & x(2,N) \\ \vdots & \vdots & \ddots & \vdots \\ x(M,1) & x(M,2) & \cdots & x(M,N) \end{bmatrix} \quad (3.23)$$

where $x(m,n) = x((m-1)(N+P)+n)$ stands for the nth sample of the mth received pilot symbol. Since the pilot contains M identical symbols, i.e.

$$s(m,n) = p(n), \quad m = 1, 2, \cdots, M, \quad n = 1, 2, \cdots, N \quad (3.24)$$

where $s(m,n)$ denotes the nth sample of the mth transmitted pilot symbol, (3.20) can be represented as

$$x(m,n) = e^{jm\Omega} \cdot \alpha(n) + e^{-jm\Omega} \cdot \beta(n) \quad (3.25)$$

where

$$\begin{aligned} \alpha(n) &= e^{j\Delta\omega T_s n} p(n) * d(n) \\ \beta(n) &= e^{-j\Delta\omega T_s n} p^*(n) * v(n) \\ \Omega &= \Delta\omega T = \Delta\omega(N+P)T_s \end{aligned} \quad (3.26)$$

and T is the pilot symbol period. Therefore, each column of X in (3.23), denoted as $x(n)$, can be expressed as a superposition of two tones as follows:

$$x(n) = \begin{bmatrix} e^{j\Omega} & e^{-j\Omega} \\ e^{j2\Omega} & e^{-j2\Omega} \\ \vdots & \vdots \\ e^{jM\Omega} & e^{-jM\Omega} \end{bmatrix} \begin{bmatrix} \alpha(n) \\ \beta(n) \end{bmatrix} \stackrel{def}{=} \Omega \begin{bmatrix} \alpha(n) \\ \beta(n) \end{bmatrix}. \quad (3.27)$$

The formulation above leads to a classic multi-line spectral estimation problem. Ω is the parameter of interest, while $\alpha(n)$ and $\beta(n)$ are treated as nuisance unknowns. The NLS method described in [28] (maximum likelihood estimator under white Gaussian noise) becomes directly applicable. Particularly in our case, the two frequencies in $x(n)$ actually contain only one variable with a different sign. As a result, a low cost one-dimension search is sufficient. The frequency estimation can thus be obtained by searching the maxima of the following:

$$\hat{\Omega} = arg \max_{\Omega} \left[tr\{ \Omega(\Omega^H \Omega)^{-1} \Omega^H \hat{R} \} \right] \quad (3.28)$$

where $tr(\cdot)$ denotes the trace of a matrix and $\hat{\mathbf{R}}$ is the sample covariance matrix

$$\hat{\mathbf{R}} = \mathbf{XX}^H. \tag{3.29}$$

Note that the algorithm proposed in [26] can be considered as a special case of (3.28) in the absence of I/Q imbalance: when $\beta(n) = 0$, (3.28) can be simplified as

$$\hat{\Omega} = arg \max_{\Omega} \left[\Omega^H \hat{\mathbf{R}} \Omega \right] \tag{3.30}$$

with $\Omega \overset{\text{def}}{=} \left[e^{j\Omega} \ e^{j2\Omega} \cdots e^{jM\Omega} \right]^T$.

Special attention should be given to Ω. In our application, Ω becomes ill-conditioned when the initial offset is close to zero, which leads to poor estimation accuracy around zero frequency. To maintain a proper condition number regardless of the initial offset, an additional $\pi/4$ rotation can be introduced to all the even pilot symbols. The resulting Ω thus becomes always well-conditioned

$$\Omega = \begin{bmatrix} e^{j\Omega} & e^{-j\Omega} \\ e^{j2\Omega} \cdot e^{j\pi/4} & e^{-j2\Omega} \cdot e^{-j\pi/4} \\ \vdots & \vdots \\ e^{jM\Omega} \cdot e^{j\pi/4} & e^{-jM\Omega} \cdot e^{-j\pi/4} \end{bmatrix}. \tag{3.31}$$

Once the compensation structure is set, the remaining task is to determine the optimum filter coefficients $w(n)$ and the phase compensator b that best remove the I/Q imbalance.

Notice from (3.25) that without I/Q imbalance ($\beta(n) = 0$), the two adjacent received symbols should only differ by a phase rotation, $e^{j\Omega} \cdot e^{\pm\pi/4}$, determined by the carrier offset and the pre-injected rotation. This phase rotation is known to the receiver after the frequency offset has been estimated. However, the presence of I/Q imbalance ($\beta(n) \neq 0$) alters this relationship between the received pilot symbols. This observation suggests a procedure to determine $w(n)$ and b based on such phase rotations. In other words, the optimum $w(n)$ and b can be estimated by minimizing the following:

$$[w(n), b]_{opt} = arg \min_{w(n), b} \left\{ \sum_{m=1}^{M-1} \sum_{n=1}^{N+L-1} |x'(m+1, n) - C_m \cdot x'(m, n)|^2 \right\} \tag{3.32}$$

where $x'(m, n) = x'_I(m, n) + j \cdot x'_Q(m, n)$ denotes the I/Q compensated signal and C_m represents the phase rotation between adjacent symbols

$$C_m = e^{j\Omega_m} = \begin{cases} e^{j\Omega} \cdot e^{j\pi/4} & : \quad m = odd \\ e^{j\Omega} \cdot e^{-j\pi/4} & : \quad m = even \end{cases} \tag{3.33}$$

Note that Ω_m or C_m is considered known by simply substituting Ω with its estimate $\hat{\Omega}$ already obtained in Section 3.3.3.

According to the compensation structure in Figure 3.10, $x_I'(m, n)$ and $x_Q'(m, n)$ are constructed as follows:

$$x_I'(m, n) = x_I(m, n) \otimes w(n) + b \cdot x_Q(m, n) \otimes \delta(n - L/2) \quad (3.34)$$

$$x_Q'(m, n) = x_Q(m, n) \otimes \delta(n - L/2). \quad (3.35)$$

Define the following matrix representations:

$$\mathbf{X}_I(m) = \begin{bmatrix} x_I(m, 1) & & & \\ x_I(m, 2) & x_I(m, 1) & & \\ \vdots & x_I(m, 2) & \ddots & \\ x_I(m, N) & \vdots & \ddots & x_I(m, 1) \\ & x_I(m, N) & \vdots & x_I(m, 2) \\ & & \ddots & \vdots \\ & & & x_I(m, N) \end{bmatrix}_{(N+L-1) \times L}$$

$$\mathbf{x}_I'(m) = [x_I'(m, 1) \; x_I'(m, 2) \cdots x_I'(m, N + L - 1)]^T \quad (3.36)$$

and use similar notations for the Q channel signal (with subscript Q). The I/Q compensation in (3.34) and (3.35) can be rewritten in matrix forms as follows:

$$\mathbf{x}_I'(m) = [\mathbf{X}_I(m) \quad \mathbf{X}_Q(m)\mathbf{1}_L] \begin{bmatrix} \mathbf{w} \\ b \end{bmatrix}$$

$$\mathbf{x}_Q'(m) = \mathbf{X}_Q(m)\mathbf{1}_L \quad (3.37)$$

where

$$\mathbf{w} = [w(1) \; w(2) \cdots w(L)]^T$$

$$\mathbf{1}_L = [\underbrace{0 \cdots 0}_{\frac{L-1}{2}} \; 1 \; \underbrace{0 \cdots 0}_{\frac{L-1}{2}}]^T. \quad (3.38)$$

Therefore, the two adjacent pilot symbols after compensation are related by

$$\begin{bmatrix} \mathbf{x}_I'(m + 1) \\ \mathbf{x}_Q'(m + 1) \end{bmatrix} = \begin{bmatrix} \cos\Omega_m & -\sin\Omega_m \\ \sin\Omega_m & \cos\Omega_m \end{bmatrix} \begin{bmatrix} \mathbf{x}_I'(m) \\ \mathbf{x}_Q'(m) \end{bmatrix} + \mathbf{N}(m), \quad (3.39)$$

which can be further expressed as

$$\mathbf{A}(m) \begin{bmatrix} \mathbf{w} \\ b \end{bmatrix} = \mathbf{B}(m) + \mathbf{N}(m) \quad (3.40)$$

where

$$\mathbf{A}(m) = \begin{bmatrix} \mathbf{X}_I(m)\cos\Omega_m - \mathbf{X}_I(m + 1) & \mathbf{X}_Q(m)\mathbf{1}_L\cos\Omega_m - \mathbf{X}_Q(m + 1)\mathbf{1}_L \\ \mathbf{X}_I(m)\sin\Omega_m & \mathbf{X}_Q(m)\mathbf{1}_L\sin\Omega_m \end{bmatrix}$$

$$\mathbf{B}(m) = \begin{bmatrix} \mathbf{X}_Q(m)\mathbf{1}_L\sin\Omega_m \\ \mathbf{X}_Q(m + 1)\mathbf{1}_L - \mathbf{X}_Q(m)\mathbf{1}_L\cos\Omega_m \end{bmatrix} \quad (3.41)$$

and $\mathbf{N}(m)$ stands for the residual error. Stacking $\mathbf{A}(m)$ and $\mathbf{B}(m)$ for all the pilot symbols, the cost function in (3.32) can be minimized by solving the linear least squares equation as follows:

$$\underbrace{\begin{bmatrix} \mathbf{A}(1) \\ \mathbf{A}(2) \\ \vdots \\ \mathbf{A}(M-1) \end{bmatrix}}_{\mathbf{A}} \begin{bmatrix} w \\ b \end{bmatrix} = \underbrace{\begin{bmatrix} \mathbf{B}(1) \\ \mathbf{B}(2) \\ \vdots \\ \mathbf{B}(M-1) \end{bmatrix}}_{\mathbf{B}} + \begin{bmatrix} \mathbf{N}(1) \\ \mathbf{N}(2) \\ \vdots \\ \mathbf{N}(M-1) \end{bmatrix}$$

The solution of optimum $w(n)$ and b is given by

$$\begin{bmatrix} w \\ b \end{bmatrix}_{opt} = \mathbf{A}^{\dagger}\mathbf{B} \tag{3.42}$$

where $(\cdot)^{\dagger}$ denotes the matrix pseudo-inverse.

Since the compensation coefficients are determined by restoring the known phase rotation between adjacent pilot symbols, it is also crucial that such rotation structure is robust with respect to the initial frequency offset. This reinforces the necessity of adding a $\pi/4$ rotation to all the even symbols, as mentioned in Section 3.3.3.

It should also be pointed out that for compensating the I/Q imbalance, $w(n)$ can be put in either the I or the Q branch. Similarly, the structure of the asymmetric phase compensator can also be flipped around, i.e., we can instead apply gain a on the Q branch and then a cross path from the I branch to the Q branch with gain b. The purpose of its current arrangement is to avoid the multiplication of $w(n)$ and b. Therefore, the cost function in (3.32) is quadratic, leading to a numerically friendly close-form solution.

Example 7 *In this example, the efficacy of the algorithm is illustrated with hardware test results. The test receiver board consists of all the key components of a direct-conversion receiver as depicted in Figure 3.7. The baseband OFDM waveform is generated by a PC and downloaded to a signal generator with an arbitrary waveform generation function. The signal generator modulates the baseband OFDM waveform to a certain RF frequency and plays it repeatedly with adjustable output power level. The OFDM modulated RF signal is then fed to the test receiver board, where it is down-converted to baseband and digitally sampled. A logic analyzer collects the resulting baseband digital signals and passes them to the PC.*

The OFDM system tested consists of 512 subcarriers with 64QAM coherent demodulation. The frame structure used in the test contains 10 pilot symbols with 16 samples each, followed by the OFDM payload symbols. The signal bandwidth is 4MHz and the RF carrier frequency is set at 1.8GHz in the experiment. Fig. 3.13(a) and Fig. 3.13(b) show the constellation of demodulated signals at

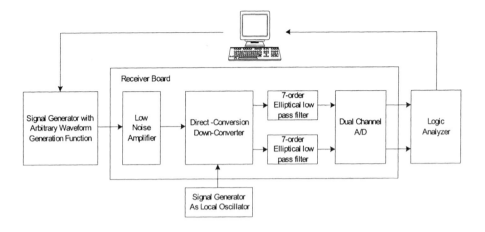

Figure 3.12: Test bench for I/Q imbalance compensation

the input power level of -60dBm. The performance improvement due to the proposed I/Q compensation can be clearly seen, as the original smeared constellation becomes much more distinguishable after the compensation. Fig. 3.14 plots the average SNR of demodulated signals vs. the signal input power. It can be observed that by employing a 5th-order FIR filter, the proposed compensation scheme provides about 4dB SNR improvement at high input signal level. However, if only the frequency independent imbalance is compensated (i.e., reduce the FIR tap to one), the performance gain reduces to 2dB.

3.4 Phase noise compensation

In general, the downconverter induced I/Q imbalance can be calibrated and compensated periodically with large intervals. The phase noise on the other hand, introduces much more rapid distortions to the signals and thus must be coped with in real time.

It is well understood that the phase noise effect on OFDM signal reception consists of two components: an inter-carrier interference (ICI) term that can be modeled as additional Gaussian noise, and a common phase error (CPE) that rotates all the subcarriers equally. While the ICI is difficult to remove due to its noise-like characteristics, the CPE can be easily corrected by estimating such rotation through continuous pilot tones embedded in OFDM symbols [30] [31]. However, existing CPE compensation schemes assume *known channel states* or at least static channels, allowing the CPE to be separated from actual channel

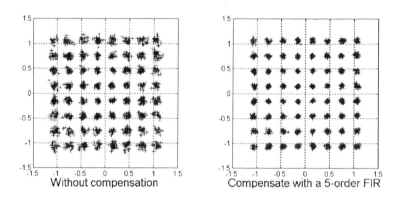

Figure 3.13: OFDM 64-QAM constellation (a) before and (b) after I/Q imbalance compensation

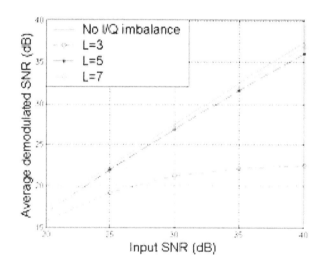

Figure 3.14: Receiver performance with I/Q compensator

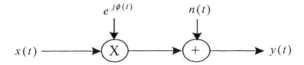

Figure 3.15: Multiplicative phase noise effect

effects. Such an assumption does not hold true for time-varying channels. Reversely, the Wiener filtering channel estimation approaches introduced in Section 3.2 may not be directly applicable in the presence of severe phase noise. Only joint consideration of CPE and channel estimation can address these problems, although in practice the two effects are often dealt with separately.

In this section, the effect of phase noise on channel estimation is first analyzed, based on which the existing time domain MMSE channel estimator is modified to accommodate the CPE. A new CPE estimator suitable for fast fading channels that requires no explicit CSI is then developed based on continuous pilot tones. The estimated CPE, together with the modified MMSE channel estimator, equalizes the composite channel, leading to reception performance close to the CPE-free case.

3.4.1 Mathematical models for phase noise

The phase noise is a multiplicative effect due to the phase variation of the oscillator. To focus on the phase noise only, we assume a simple model in Figure 3.15 where there is no center frequency variation or amplitude fluctuation in the receiver's local oscillator. Therefore, the complex representation of phase noise is given by

$$\eta(t) = e^{j\phi(t)} \simeq 1 + k \cdot \phi(t)$$

The approximation above holds when the phase variation $\phi(t)$ is small, which is the case in most practical applications. $\phi(t)$ is usually modeled as a Gaussian noise with power spectrum density (PSD) $L_\phi(f)$. There are a number of phase noise PSD models in literature [30][33]. Here we choose the one (used in the European dTTb project) that specifies the key parameters of a PLL tracked oscillator [30].

$$L_\phi(f) = 10^{-c} + \begin{cases} 10^{-a} & |f| \leq f_1 \\ 10^{-(|f|-f_1)\cdot\frac{b}{f_2-f_1}-a} & |f| \geq f_1 \end{cases}$$

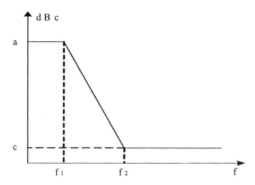

Figure 3.16: Phase noise power spectrum density

As depicted in Figure 3.16, the parameter c determines the PLL noise level, a and f_1 specify the PLL's loop characteristics, while b and f_2 describe how fast the PSD rolls off from the center frequency and where the PSD hits the noise floor, respectively. When an OFDM signal is down-converted by a non-ideal local oscillator with phase noise, the resulting frequency domain signal representation is given by [31]

$$y(k,m) \;=\; y(k,m)h(k,m)I_m(0) \tag{3.43}$$

$$+ \sum_{l=0; l \neq k}^{N-1} x(l,m)h(l,m)I_{m(l-k)} + v(k,m) \tag{3.44}$$

$$\approx \; x(k,m)h(k,m) \cdot e^{j\Phi_m} + v_{ICI}(k,m) + v(k,m)$$

where k and m are the subcarrier index and the symbol index and , respectively. $x(k,m)$, $y(k,m)$ and $h(k,m)$ denote the transmitted signal, the received signal, and the channel response, respectively. $v(k,m)$ is the additive white Gaussian noise. $I_m(l)$ represents the phase noise effect on the OFDM signal reception:

$$I_m(l) = \frac{1}{N} \sum_{l=0}^{N-1} e^{j2\pi nl/N} e^{j\phi_m(n)}$$

As shown in [30][33], the phase noise effect is composed of two parts: (i) a common phase error (CPE) $e^{j\phi_m}$, that rotates all the subcarriers equally (the amplitude variation can be ignored when $e^{j\phi_m}$ is small); and (ii) an inter-carrier interference (ICI) term, v_{ICI}, that is usually modeled as additional Gaussian noise, provided that N is sufficiently large and $x(k,m)$ are mutually independent. Both the CPE and the ICI are shown to be Gaussian variables with variance given as follows [30][33]:

$$\sigma_{CPE}^2 = \int_{-B/2}^{B/2} L_\phi(f) \cdot \sin c^2(\pi f T_{symbol}) df$$

$$\sigma_{ICI}^2 = \int_{-B/2}^{B/2} L_\phi(f) df - \sigma_{CPE}^2$$

where B is the signal bandwidth and T_{symbol} denotes the OFDM symbol period. Notice from the above expressions that given a certain phase noise spectrum $L_\phi(f)$, there is a trade-off between the CPE and the ICI (by adjusting subcarrier spacing $\frac{1}{T_{symbol}}$). Since the ICI is more difficult to compensate, OFDM system parameters are usually designed in such a way that phase-noise-induced ICI is several dB lower than the operational noise level [29]. For this reason, we will ignore the effect of ICI and only focus on the CPE in the ensuing discussion.

3.4.2   CPE estimation with channel state information

Let us first examine how the phase-noise-induced CPE affects channel estimation in both frequency and time domain filtering.

- In the frequency domain, the CPE adds a phase rotation common to all the subcarriers. Such a rotation presents a constant ambiguity on the actual channel and therefore has no impact on the frequency domain channel estimation/filtering.

- In the time domain, the CPE varies from symbol to symbol with a weak correlation between symbols [30]. As a result, the composite channel $h'(k, m) = h(k, m) \cdot e^{j\Phi_m}$ loses its original low-pass characteristics (determined by the channel Doppler spectrum), which prohibits direct application of the existing time-domain channel estimation/filtering.

As mentioned in earlier sections, pilot OFDM symbols are usually available in practical systems for channel estimation purposes. The samples in the time-frequency grid (various pilot patterns have been considered [34]) allow channel estimation using 2-D or 1+1 D Wiener filtering processes. The same pilots can be utilized for phase noise compensation. In particular, we consider the scenarios with uniformly scattered pilots and continuous pilot tones (over time) exemplified in Figure 3.17. Similar pilot patterns are employed in the DVB-T, IEEE 802.11a/g and IEEE802.16 standards.

After removing the known modulation $x(k, m)$ at pilot locations in (3.43), OFDM channel response in the presence of phase noise can be formulated similar to (3.9) as follows:

$$
\begin{aligned}
r(k, m) &= h(k, m) \cdot e^{j\Phi_m} + v(k, m) \qquad (3.45) \\
&= h'(k, m) + v(k, m)
\end{aligned}
$$

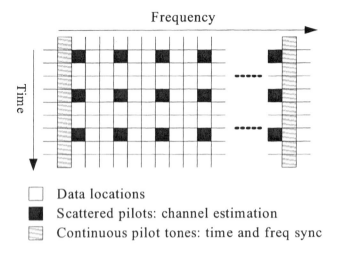

Figure 3.17: Continous and scattered pilots for CPE estimation

where $(k, m) \in \mathcal{P}$ and $\mathcal{P}$ represents the set of pilot locations. $h(k, m), h'(k, m)$ and $r(k, m)$ are the actual channel response, the composite channel response (with CPE), and the channel observation, respectively. $v(k, m)$ now represents white Gaussian noise that incorporates both thermal noise and the phase-noise-induced ICI.

From Section 3.2, channel estimation is essentially an interpolation or filtering of channel responses based on the noisy observations $r(k, m)$ at $(k, m) \in \mathcal{P}$. Unlike the phase-noise free case, we are now dealing with the composite channel $h'(k, m) = h(k, m) \cdot e^{j\Phi_m}$. Therefore, the CPE effect needs to be taken into account for coherent demodulation.

CPE estimation methods based on continuous pilot tones and *known* channel state information have been proposed in [30][31]. In particular, the CPE is estimated as

$$\widehat{\Phi}_m = \angle \left\{ \sum_{k \subset \mathcal{C}} r(k, m) h^*(k, m) \right\} \tag{3.46}$$

where $\mathcal{C}$ is the set of subcarrier indices of continuous pilot tones. In a time varying channel where the exact channel information is not available, the performance of CPE estimation in (3.46) degrades dramatically. This indicates that for rapid dispersive fading channels the CPE effect and the *time domain* channel response must be estimated jointly.

3.4.3 Time domain channel estimation in the presence of CPE

Since the frequency domain 1-D channel estimation/filtering is not affected by the presence of CPE, one can deal with such tangled effects in two steps. First, we modify the time domain MMSE channel estimator described in Section 3.2.3 to accommodate the CPE effect for composite channel estimation. Then, we develop a CPE estimator that requires no explicit CSI by exploiting continuous pilot tones.

Since CPE only affects time domain filtering, in the ensuing discussion we consider channel estimation at a given subcarrier and drop the frequency index k in (3.45) unless explicitly indicated otherwise.

From Section 3.2.3 on 1-D time domain filtering, we introduce the new channel matrices to take into account the CPE effect as follows,

$$\mathbf{h}' = \mathbf{\Phi}\mathbf{h}$$
$$\mathbf{h}'_{\mathcal{P}_t} = \mathbf{\Phi}_{\mathcal{P}_t}\mathbf{h}_{\mathcal{P}_t}$$

where

$$\mathbf{\Phi} = diag\left\{ e^{j\Phi(1)}, e^{j\Phi(2)}, \cdots e^{j\Phi(N)} \right\}$$
$$\mathbf{\Phi}_{\mathcal{P}_t} = diag\left\{ e^{j\Phi(\mathcal{P}_1)}, e^{j\Phi(\mathcal{P}_2)}, \cdots e^{j\Phi(|\mathcal{P}_t|)} \right\}$$

Replacing the correlation matrices in (3.11) with the covariance matrices of the composite channel vectors, the MMSE estimate of the composite channel in the presence of CPE needs to be modified to

$$\widehat{\mathbf{h}}_{MMSE} = \mathbf{\Phi}\mathbf{F}_t\mathbf{\Phi}_{\mathcal{P}_t}^H \mathbf{r}_{\mathcal{P}_t} \tag{3.47}$$

where $\mathbf{F}_t$ is defined in (3.11)

The above estimator makes intuitive sense, since it actually suggests that the CPE be first removed from observations at pilot locations before Wiener filtering is applied to smooth out noise. The CPE is then compensated on all the data symbols to obtain the composite channel estimation.

3.4.4 CPE estimation without explicit CSI

Now the problem reduces to determining the CPE without explicit knowledge of the channel information.

Parameters	Specifications
Sampling Frequency (Fs)	4MHz
FFT size	512
Cyclic prefix	64
Scattered pilot spacing (frequency)	3
Scattered pilot spacing (time)	3
# of continuous pilot tones	16
Truncating length of frequency domain MMSE	64
Time domain MMSE block size	20

Table 3.1: OFDM system parameters in joint CPE and channel estimation

Due to its weak time correlation, the CPE must be estimated symbol-by-symbol based on the continuous pilot tones (available in all OFDM symbols). Unlike the approach in (3.46) [30][31], the method described here does not require explicit channel state information. In [35], it is shown that $\mathbf{\Phi}$ can be estimated by minimizing the following:

$$\Phi_0 = \arg \min_{\Phi} E \left\{ |\mathbf{\Phi W \Phi}^H \mathbf{R}_\mathcal{C} - \mathbf{R}_c| \right\} \tag{3.48}$$

where

$$\mathbf{R}_\mathcal{C} = \begin{bmatrix} r(\mathcal{C}_1, 1) & r(\mathcal{C}_2, 1) & \cdots & r(|\mathcal{C}|, 1) \\ r(\mathcal{C}_1, 2) & r(\mathcal{C}_2, 2) & \cdots & r(|\mathcal{C}|, 2) \\ \vdots & \vdots & \ddots & \vdots \\ r(\mathcal{C}_1, N) & r(\mathcal{C}_2, N) & \cdots & r(|\mathcal{C}|, N) \end{bmatrix}$$

contains observations at the continuous pilot location $\mathcal{C}$ over N OFDM symbol periods;

$$\mathbf{W} = \mathbf{R_{hh}} \left(\mathbf{R_{hh}} + \frac{\mathbf{I}}{SNR} \right)^{-1}$$

is a function of the time channel vector covariance matrix $\mathbf{R_{hh}}$.

Example 8 *In this example, the joint CPE and channel estimator is applied to an OFDM system with parameters listed in Table 3.1. The pilot pattern shown in Figure 3.17 is assumed. The UMTS channel (COST207-TU) with different Doppler frequencies is used. The Doppler frequency is fixed at $F_d T_s = 0.05$. We choose the phase noise parameters as suggested in [30] : $a = 6, c = 10.5, f_1 = 1kHz$, and $f_{2=10kHz}$. The MSE of composite channel estimates is used as the performance measure for comparing the following 3 different schemes in rapid dispersive fading channels with phase noise:*

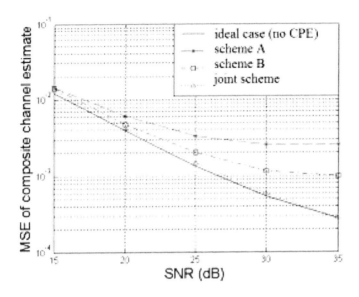

Figure 3.18: Performance of joint phase noise and channel estimation

- *Scheme A: channel estimation with time-domain Wiener filtering as in (3.11) without considering the phase noise effect*

- *Scheme B: channel estimation with time-domain Wiener filtering as in (3.11), followed by the existing CPE estimation as in (3.46)*

- *Joint Scheme: CPE estimation using (3.48), followed by the modified time-domain Wiener filtering as in (3.47)*

From Figure 3.18, it is clear that the presence of CPE causes an irreducible error floor in Schemes A and B. The CPE compensation approach in B does alleviate the CPE effect, however the error floor is only reduced but not eliminated. In contrast, Scheme C with joint CPE and channel estimation performs almost as well as that without CPE.

3.5   Summary

In this chapter, we address OFDM pre- and post- demodulation issues emerging from system imperfections and channel dispersion. Problems related to carrier frequency offset, time- and frequency-selective channels, I/Q imbalance, and phase noise, are formulated. For each problem, we present solutions and algorithms that are practical and efficient. Simulations and application examples are included to show how these techniques can be employed in real OFDM systems.

Bibliography

[1] P. Tan and N. C. Beaulieu, "Reduced ICI in OFDM systems using the better than raised-cosine pulse," *IEEE Commun. Lett.*, vol. 8, no. 3, March 2004, pp 135-137.

[2] T. M. Schmidl and D. C. Cox, "Robust frequency and timing synchronization for OFDM," *IEEE Trans. Commun.*, vol. 45, pp. 1613-1621, Dec. 1997.

[3] J. van de Beek, M. Sandell, and P. O. Oorjesson, "ML estimation of time and frequency offset in OFDM systems," *IEEE Trnas. Signal Processing*, vol. 45, pp. 1800-1805, July 1997.

[4] T. Pollet, M. V. Bladel and M. Moeneclaey, "BER sensitivity of OFDM systems to carrier frequency offset and Wiener phase noise," *IEEE Trans. Communications*, vol. 43, pp. 191-193, Feb./Mar./Apr. 1995.

[5] H. Liu and U. Tureli, "A high-efficiency carrier estimator for OFDM communications," *IEEE Commun. Lett.*, vol. 2, no. 4, April 1998, pp 104-106.

[6] R. O. Schmidt, "Multiple emitter location and signal parameter estimation," in *Proc. RADC Spectral Estimation Workshop*, Griffiss Air Force Base, NY, pp. 243-258, 1979.

[7] B. Chen and H. Wang, "Blind estimation of OFDM carrier frequency offset via oversampling," *IEEE Trans. on Signal Processing*, vol 52, no. 7, July 2004, pp 2047 - 2057.

[8] X. Ma and G. B. Giannakis, "Unifying and optimizing null-subcarrier based frequency-offset estimators for OFDM," in *Proc. Int. Conf. Inform, Commun., Signal Processing*, Singapore, Oct. 2001.

[9] B. Chen, "Maximum likelihood estimation of OFDM carrier frequency offset," *IEEE Signal Processing Lett.*, vol.9, pp 123-126, April 2002.

[10] U. Tureli, P.Honan and H. Liu, "Low Complexity Nonlinear Least Squares Carrier Offset Estimator for OFDM: Identifiability, Diversity and Performance," *IEEE Trans. Signal Processing*, 52(9):2441 –2452, September 2004.

[11] S. Haykin, *Adaptive Filter Theory*, second edition, Prentice-Hall, Englewood Cliffs, NJ, 1991.

[12] R. Negi and J. Cioffi, "Pilot tone selection for channel estimation in a mobile OFDM systems," *IEEE Trans. Consumer Electronics*, vol. 44, no. 3, Aug. 1998.

[13] J. van de Beek, O. Edfors, M. Sandell, S. K. Wilson, and P. O. Borjesson, "On channel estimation in OFDM systems," *Proc. IEEE 45th Vehicular Technology Conference*, Chicago, IL, pp. 815-819, July 1995.

[14] O. Edfors, M. Sandell, J. J. van de Beek, S. K. Wilson, and P. O. Borjesson, "OFDM channel estimation by singular value decomposition," *IEEE Trans. Communications*, vol. 46, pp 931-939, July 1998.

[15] T.A. Chen, Michael Fitz, Shengchao Li, M.D. Zoltowski, "Two-dimensional space-time pilot-symbol assisted demodulation for frequency-nonselective Rayleigh fading channels," *IEEE Transactions on Communications*, vol. 52, issue: 6 , pp 953 - 963, June 2004.

[16] Yang et al, "Channel estimation for OFDM transmission in multipath fading channels based on parametric channel modeling," *IEEE Trans. Communications*, 49(3), pp. 467-479, March 2001.

[17] Y. Li, L. J. Cimini, and N. R. Sollenberger, "Robust channel estimation for OFDM systems with rapid dispersive fading channels," *IEEE Trans. Communications*, vol. 46, pp. 902-915, July 1998.

[18] H. Yaghoobi, "Scalable OFDMA physical layer in IEEE 802.16 wireless-MAN," *Intel Technology Journal*, vol. 8, issue 3, 2004.

[19] M. Valkama, M. Renfors, V. Koivunen, "Advanced Methods for I/Q Imbalance Compensation in Communication Receivers", *IEEE Trans. Signal Processing*, Vol. 49, No. 10, Oct 2001, pp.2235-2344.

[20] K. Pun, et al, "Correction of Frequency-dependent I/Q Mismatches in Quadrature Receivers", *Electronics Letters*, Volume: 37, Issue 3, 8 Nov. 2001 Page(s): 1415 -1417.

[21] A. Schuchert, R. Hasholzner, P. Antoine, "A novel IQ imbalance compensation scheme for the reception of OFDM signals", *IEEE Transactions on Consumer Electronics*, Volume: 47 Issue: 3 , Aug. 2001, Page(s): 313 -318.

[22] S. Simoens, et al, "New I/Q imbalance modeling and compensation in OFDM systems with frequency offset", *PIMRC*, 2002. The 13th IEEE International Symposium on Personal, Indoor and Radio Communications, Volume: 2, 2002 Page(s): 561 -566.

[23] R.V. Nee and R. Prasad, *OFDM for Wireless Multimedia Communications*, Artech House Publishers, 2000.

[24] M. Valkama, M. Renfors and V. Koivunen, "Compensation of frequency-selective I/Q imbalances in wideband receivers: models and algorithms," in *Proc. IEEE SPAWC'2001*, pp. 42-45, Taoyuan, Taiwan, Mar. 2001.

[25] J. K. Cavers and M. W. Liao, "Adaptive compensation for imbalance and offset losses in direct conversion transceivers," *IEEE Trans. Vehicular Tech.*, vol. 42, pp. 581-588, Nov. 1993.

[26] J. Li, G. Liu and G. B. Giannakis, "'Carrier frequency offset estimation for OFDM-based WLANs," *IEEE Trans. Signal Processing*, vol. 8, pp. 80-82, Mar. 2001.

[27] V. K. P. Ma and T. Ylamurto, "Analysis of IQ imbalance on initial frequency offset estimation in direct down-conversion receivers," in *Proc. IEEE SPAWC'2001*, pp. 158-161, Taoyuan, Taiwan, Mar. 2001.

[28] P. Stoica and R. Moses, *Introduction to Spectral Analysis*, Prentice-Hall, Englewood Cliffs, NJ, 1997.

[29] M. Septh, S. A. Fechtel, G. Fock and H. Meyr, "Optimum receiver design for wireless broad-band systems using OFDM – Part I," *IEEE Trans. Commun.*, vol. 47, pp. 1668-1677, Nov. 1999.

[30] P. Robertson and S. Kaiser, "Analysis of the effects of phase noise in orthogonal frequency division multiplexing (OFDM) systems," in *IEEE Proc. ICC'95*, vol. 3, Seattle, WA, 1995, pp. 1652-1657.

[31] S. Wu and Y. Bar-Ness, "A phase noise suppression algorithm for OFDM-based WLANs," *IEEE Commun. Letters*, vol. 6, pp. 535-537, Dec. 2002.

[32] A. G. Armada, "Understanding the effects of phase noise in orthogonal frequency division multiplexing (OFDM)," *IEEE Trans. Broadcast.*, vol. 47, pp. 153-159, June 2001.

[33] M. S. El-Tanany, Y. Wu and L. Hazy, "Analytical modeling and simulation of phase noise interference in OFDM-based digital television terrestrial broadcasting systems," *IEEE Trans. Broadcast.*, vol. 47, pp. 20-31, Mar. 2001.

[34] P. Hoeher, S. Kaiser and P. Robertson, "Two-dimensional pilot-symbol-aided channel estimation by Wiener filtering," *Proc. IEEE ICASSP'97*, vol. 3, Munich, Germany, Apr. 1997, pp. 1845-1848.

[35] G. Xing, M. Shen, and H. Liu, "Common phease error (CPE) and channel estimation for OFDM systems," in *Proc. CISS'03*, Baltimore, March 2003.

Chapter 4

PHY Layer Issues – Spatial Processing

A wireless modem equipped with multiple antennas is capable of exploiting the spatial diversity of wireless channels. The *space domain* enabled by the antenna array introduces an additional dimension to the time-frequency radio resource. Through space-time processing, an antenna array can improve key operational parameters such as the SINR, the data rate, and the outage probability over a single antenna wireless system. Other major benefits of spatial processing include

- better interference/jammer rejection
- higher power efficiency
- lower cost, distributed hardware

Antenna array technology encompasses a wide variety of techniques that can be used at both the base-station and in the user terminal. There are two main classes of spatial processing techniques, namely, *beamforming* and *space-time coding*. Early beamforming techniques are *SINR-oriented*, involving linear combining of transmit and received signals from multiple antennas. The main objective is to maximize the instantaneous signal-to-interfernece-and-noise ratio (SINR). More recent techniques, such as space-time coding, are mostly *diversity-and-efficiency-oriented*, focusing on lowering the BER of a high-speed link over fading channels. As will be shown in the ensuing discussion, these two types of techniques can be synergistically combined for broadband OFDM systems.

4.1 Antenna array fundamentals

To better understand the principles of array processing, consider the scenario depicted in Figure 4.1 where a narrowband signal $s(t)e^{j\omega_0 t}$ impinges on an array

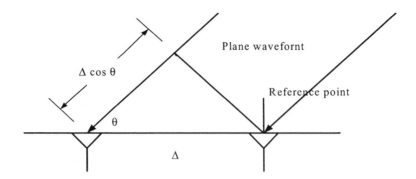

Figure 4.1: A 2-element array and its response to narrowband signal

of two elements from angle θ. The carrier frequency of the narrow band signal is ω_0, and the spacing between the two antenna elements is Δ.

From the figure, it can be seen that the propagation delay between the first and second antenna elements is

$$\tau = \frac{\Delta \cos \theta}{c}$$

where c is the speed of light. $\omega_0 = 2\pi c/\lambda$. Consequently, the carrier phase shift between the two antenna element is given by

$$\phi = \omega_0 \tau = \omega_0 \frac{\Delta \cos \theta}{c} = \frac{2\pi \Delta \cos \theta}{\lambda}.$$

Putting the two *baseband* array outputs in vector form, we have

$$\begin{aligned} \mathbf{y}(t) &= \begin{bmatrix} y_1(t) & y_2(t) \end{bmatrix}^T = \begin{bmatrix} s(t) & s(t-\tau) \end{bmatrix}^T \qquad (4.1) \\ &= \begin{bmatrix} 1 & e^{-j\frac{2\pi\Delta \cos\theta}{\lambda}} \end{bmatrix}^T s(t) = \mathbf{a}(\theta)s(t). \end{aligned}$$

An important assumption we invoked here is that the signal is *narrow-band* and therefore the propagation delay between antennas, τ, is insignificant relative to the coherence time of the signal. In other words, $s(t)$ can be regarded as a constant over the period τ. (4.1) can be easily extended to cases with M ($M > 2$) receiving antennas:

$$\begin{aligned} \mathbf{y}(t) &= \begin{bmatrix} 1 & e^{-j\frac{2\pi\Delta \cos\theta}{\lambda}} & \cdots & e^{-j\frac{2\pi\Delta(M-1)\cos\theta}{\lambda}} \end{bmatrix}^T s(t) \qquad (4.2) \\ &= \mathbf{a}(\theta)s(t). \end{aligned}$$

The above $\mathbf{a}(\theta)$ is termed as the *steering vector* - it is a function of the angle-of-arrival (AOA) and the array configuration. With only one path between the source (e.g., the user) and the array (e.g., the base-station), the array response vector is identical to the steering vector up to a complex scalar.

In most practical situations with multipath reflections, the antenna output vector is the superposition of these multipath signals:

$$\mathbf{y}(t) = \sum_{l=1}^{L} \nu_l \mathbf{a}(\theta_l) s(t) = \mathbf{a} s(t) \tag{4.3}$$

where L denotes the total number of coherent paths, and ν_l represents the complex gain from direction θ_l. The composite channel response, $\mathbf{a}$, termed as the *spatial signature* (SS), characterizes the spatial propagation channel between the transmitter and the receiver antenna array.

For a multiuser system with additional noise, the array output can be written as

$$
\begin{aligned}
\mathbf{y}(t) &= \sum_{i=1}^{P} \mathbf{a}_i s_i(t) + \mathbf{n}(t) \tag{4.4}\\
&= \mathbf{A}\mathbf{s}(t) + \mathbf{n}(t)\\
\mathbf{A} &= \begin{bmatrix} \mathbf{a}_1 & \mathbf{a}_2 & \cdots & \mathbf{a}_P \end{bmatrix}\\
\mathbf{s}(t) &= \begin{bmatrix} s_1(t) & s_2(t) & \cdots & s_P(t) \end{bmatrix}^T
\end{aligned}
$$

where P is the number of users, $\mathbf{n}(t)$ is the noise vector, and $\mathbf{A}$ is defined as the *array manifold* whose columns are the spatial signatures.

For *broadband* signals where the delays between multipaths are not negligible, the memoryless spatial signature vector $\{\mathbf{a}_i\}$ must be replaced with vector FIR filters $\{\mathbf{h}_i(t)\}$, yielding the following input-output relation:

$$\mathbf{y}(t) = \sum_{i=1}^{P} \mathbf{h}_i(t) * s_i(t) + \mathbf{n}(t) \tag{4.5}$$

To utilize conventional spatial processing techniques developed for narrowband applications, one can apply OFDM to convert the broadband channel into N parallel narrowband subchannels:

$$\mathbf{y}(k,t) = \sum_{i=1}^{P} \mathbf{a}_{k,i} s_i(k,t) + \mathbf{n}(k,t), \, k = 1, ..., N \tag{4.6}$$

where k is the subcarrier index, and $\mathbf{a}_{k,i}$ is the spatial signature of the ith user at subcarrier frequency k. As a result, parallel implementation of narrowband spatial processing algorithms becomes readily applicable on individual subcarriers.

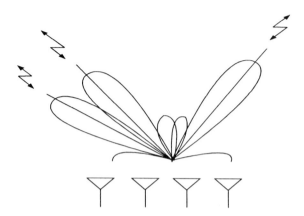

Figure 4.2: Antenna array and beamforming

4.2   Beamforming

Antenna array beamforming has been an active topic for wireless communications for the last 30 years. Beamforming is a linear space-time process that enhances the signal-of-interest through *weight-and-sum* operations. Beamforming can be performed both in transmission (TX) and reception (RX).

Consider the RX array output expressed in Equation (4.4). The signal-of-interest, say, $s_i(t)$, is subject to both interference and noise. A beamformer can extract $s_i(t)$ from $\mathbf{y}(t)$ by properly combining the M antenna outputs with a set of weights $\mathbf{w}_i = [w_{i,1}, w_{i,2}, \cdots, w_{i,M}]^T$:

$$
\begin{aligned}
\widehat{s}_i(t) &= \mathbf{w}^H \mathbf{y} = \sum_{m=1}^{M} w_{i,m}^* y_m(t) \tag{4.7} \\
&= \mathbf{w}_i^H \mathbf{a}_i s_i(t) + \sum_{j=1, j \neq i}^{P} \mathbf{w}_i^H \mathbf{a}_j s_j(t) + \mathbf{w}_i^H \mathbf{n}(t).
\end{aligned}
$$

A few representative beamformers are described here based on the way a weight vector is constructed.

4.2.1   Coherent combining

The coherent combiner is essentially a "single-user" beamformer in the sense that it performs beamforming solely based on $\mathbf{a}_i$ without regard to the interference or noise characteristics.

$$\mathbf{w}_i = \mathbf{a}_i$$

$$\widehat{s}_i(t) = ||\mathbf{a}_i||^2 s_i(t) + \sum_{k=1,k\neq i}^{K} \mathbf{a}_i^H \mathbf{a}_k s_k(t) + \mathbf{a}_i^H \mathbf{n}(t).$$

While enhancing the SINR of the signal-of-interest, the coherent combiner also suppresses interference since the spatial signatures are independent in general. The scheme is effective in the presence of a large number of interfering signals.

4.2.2 Zero-forcing

In the presence of a limited number of dominant interfering signals, a more effective scheme is the zero-forcing beamformer. Instead of suppressing the interference, a zero-forcing beamformer cancels out interference by putting a null in the direction defined by its spatial signature.

$$\mathbf{w}_i \quad : \quad \mathbf{w}_i^H \mathbf{a}_j = \delta_{i,j}$$

$$\widehat{s}_i(t) = s_i(t) + \mathbf{w}_i^H \mathbf{n}(t)$$

Given an array with M elements, the maximum number of interfers that can be eliminated is $M-1$.

4.2.3 MMSE reception (optimum linear receiver)

The zero-forcing beamformer, while eliminating the interference, may at the same time amplify the noise due to its sidelobes. The "optimum" beamformer is defined as the one that minimizes the mean-squared error (MMSE) between the transmitted signal and the beamformer output:

$$\mathbf{w}_{i,mmse} = \arg_{\mathbf{w}_i} \min E\{\widehat{s}_i(t) - s_i(t)\}$$

Using the orthogonality principle [2], one can easily obtain the MMSE beamformer as follows:

$$\mathbf{w}_{i,mmse} = \mathbf{R_y}^{-1} \mathbf{a}_i$$

$$MSE_i = 1 - \mathbf{a}_i^H \mathbf{R_y}^{-1} \mathbf{a}_i$$

$$SINR_i = \frac{|\mathbf{a}_i^H \mathbf{R_y}^{-1} \mathbf{a}_i|^2}{|\sum_{j=1,j\neq i}^{M} \mathbf{a}_i^H \mathbf{R_y}^{-1} \mathbf{a}_j|^2 + |\mathbf{a}_i^H \mathbf{R_y}^{-1} \mathbf{a}_i|^2 \sigma_n^2}$$

where $\mathbf{R_y} = E\{\mathbf{y}\mathbf{y}^H\}$ is the covariance matrix of the array output.

The most obvious benefit of beamforming is the SINR gain, as shown in the following propositions. Another key benefit is the increase in diversity order, which is critical in fading channels. More discussion will be provided in later Sections.

Proposition 1 *For RX beamforming, the average SNR increases linearly with respect to the number of receivers.*

Proof. In RX beamforming,

$$\mathbf{y} = \mathbf{a}s + \mathbf{n}$$

where

$$\mathbf{n} \sim N(\mathbf{0}, \sigma_n^2 \mathbf{I}); \quad var(s) = \sigma_s^2$$

Applying the matched filter theorem, the optimum beamformer is given by

$$\begin{aligned} \mathbf{w} &= \alpha \mathbf{a} \\ \widehat{s} &= \mathbf{w}^H \mathbf{y} \end{aligned} \tag{4.8}$$

Assuming that the channel elements $\{a_m\}$ are i.i.d. with variance σ_a^2, and the resulting SNR is thus

$$SNR = \frac{\|\mathbf{a}\|^2}{\sigma_n^2}\sigma_s^2 \Rightarrow E\{SNR\} = M\frac{\sigma_s^2 \sigma_a^2}{\sigma_n^2}$$

∎

It is important to realize that (i) the linear increase in SNR does not necessarily transfer into a linear data rate increase due to the diminishing return in achieveable data rate; and (ii) RX beamforming also provides an increase in diversity order to M which is important in fading channels.

Transmit beamforming can be carried out in a similar fashion. Consider the case with M transmitters and 1 receiver, the TX beamformer weights the signal $s(t)$ with an $M \times 1$ vector before delivering them over the M transmitters:

$$\mathbf{x}(t) = \mathbf{w}s(t)$$

The receiver output is the superposition of the M copies of the original signal, attenuated by the channel coefficients:

$$y = \sum_{i=1}^{M} a_i^* x_i(t) = \mathbf{a}^H \mathbf{w}s(t) \tag{4.9}$$

The duality between receive beamforming (4.8) and transmit beamforming (4.9) is obvious.

Proposition 2 *For TX beamforming, the average SNR increases linearly with respect to the number of transmitters.*

Proof. In TX beamforming, the transmitted signal is

$$\mathbf{x} = \mathbf{w}s$$

whereas the received signal is a scalar

$$y = \mathbf{a}^H \mathbf{w}s + n \tag{4.10}$$

The optimum TX beamformer, subject to the total power constraint $\|\mathbf{w}\| = 1$, is

$$\mathbf{w} = \mathbf{a}/\|\mathbf{a}\|,$$

and the resulting SNR is given by

$$SNR = \frac{\|\mathbf{a}\|^2}{\sigma_n^2}\sigma_s^2 \Rightarrow E\{SNR\} = M\frac{\sigma_s^2\sigma_a^2}{\sigma_n^2} \tag{4.11}$$

∎

In many practical situations, the increase of transmit antennas also means a linear increase in total powers. In this case, the average SNR will increase quadratically with respect to the number of transmitters.

4.2.4 SDMA

Space-division multiple-access augments regular multiple-access schemes such as TDMA and CDMA by accommodating multiple users (signals) in one radio resource unit (time or frequency, or code). The idea is to multiplex signals spatially using beamforming or other more aggressive multiuser detection methods. Conceptually, SDMA can *multiply* the network capacity without additional radio resources. In reality however, the ability to separate co-channel signals depends critically on the spatial characteristics of the multiple users.

In RX SDMA, a commonly used multiuser detector is the beamformer. Using the result in (4.7), signal separation can be accomplished by

$$\hat{\mathbf{s}}(t) = \begin{bmatrix} \hat{s}_1(t) \\ \vdots \\ \hat{s}_P(t) \end{bmatrix} = \begin{bmatrix} \mathbf{w}_1^H \mathbf{y}(t) \\ \vdots \\ \mathbf{w}_P^H \mathbf{y}(t) \end{bmatrix} = \mathbf{W}^H \mathbf{y}(t).$$

The beamforming matrix $\mathbf{W}$ can be constructed based on any of the aforementioned principles. Otherwise nonlinear detection, e.g., the maximum likelihood detection can be employed to recover the multiple symbol streams:

$$\hat{\mathbf{s}}(t) = \begin{bmatrix} \hat{s}_1(t) \\ \vdots \\ \hat{s}_P(t) \end{bmatrix}^T = \arg_{s_i(t) \in \mathcal{S}} \min \|\mathbf{y}(t) - \sum_{k=1}^{P} \mathbf{a}_k s_k(t)\|^2$$

where $\mathcal{S}$ is the finite set of all possible symbol waveforms.

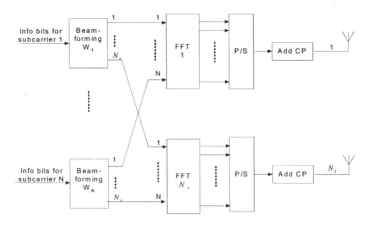

Figure 4.3: Broadband beamforming with OFDM

4.2.5   Broadband beamforming

If the signals are broadband as modeled in (4.5), the conventional narrowband beamforming techniques are not directly applicable.

Fortunately with OFDM, the broadband channel can be converted into parallel narrowband subchannels as in (4.6). Accordingly, narrowband beamforming can be implemented on individual subchannels. Figure 4.3 depicts the general block diagram of OFDM TX beamforming. The OFDM RX beamformers can be similarly constructed on subcarriers. Notice that since the broadband channels are *frequency selective*, the spatial channel characteristics vary from subcarrier to subcarrier. By default, different beamforming vectors will be needed on different subchannels. In practical situations, further channel structure information can be exploited to simplify the spatial operations - see ensuring sections for more discussions.

As a final note on beamforming, it must be pointed out that in practice the TX beamforming is significantly more challenging than RX beamforming for the following reasons:

- *Difficulty in obtaining the TX spatial signature*: the RX spatial signature in (4.3) and the TX spatial signature (4.10) are both time and frequency dependent, and therefore different in general. While the RX spatial signature, $\mathbf{a}_R$, can be estimated along with the received signals, the TX spatial signature, $\mathbf{a}_T$, can only be obtained indirectly. In a TDD (time-division multiplexing) system for example, one can estimate $\mathbf{a}_T$ as $\mathbf{a}_T = \mathbf{a}_R$. However such an assumption is only valid if the TDD time frames are short

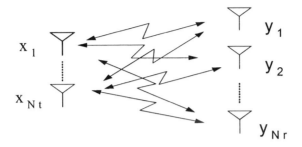

Figure 4.4: MIMO channel illustration

relative to the coherent time of the channels. This approach may be problematic in asymmetric data system where the availability of RX signals is not guaranteed.

- *Hardware induced phase ambiguity*: Both (4.3) and (4.10) ignore the fact that practical transceiver hardware is not perfectly balanced or aligned. In reality the *composite RX and TX* spatial signatures are given by

$$\mathbf{g}_R \odot \mathbf{a}_R = \begin{bmatrix} g_R(1)a_R(1) & g_R(2)a_R(2) & \cdots & g_R(M)a_R(M) \end{bmatrix}^T$$
$$\mathbf{g}_T \odot \mathbf{a}_T = \begin{bmatrix} g_T(1)a_T(1) & g_T(2)a_T(2) & \cdots & g_T(M)a_T(M) \end{bmatrix}^T$$

respectively, where $\mathbf{g}_R$ is the complex gain vector of the receiver hardware chains and $\mathbf{g}_T$ is the complex gain vector of the transmit hardware chains. In most RX processing, only the composite spatial signature is needed. Such is not the case in TX beamforming where $\mathbf{g}_R$, or at least $\{g_T(m)/g_R(m)\}$, must be calibrated and compensated.

- *Broadband channels:* the calibration problem is further complicated by the fact that both $[\mathbf{a}_R, \mathbf{a}_T]$ and $[\mathbf{g}_R, \mathbf{g}_T]$ are frequency dependent in broadband systems.

4.3 MIMO channels and capacity

Traditional beamforming assumes multiple antennas at either the transmitter or the receiver side. When antenna arrays are available at both the transmitter and the receiver as shown in Figure 4.4, they create a channel function that has multiple-input and multiple-output (MIMO) . The MIMO architecture has inspired much breakthrough research since 1996 [3]. In the following, we examine the potential of MIMO channels by evaluating its capacity.

From (4.3), a MIMO channel with N_t transmitters and N_r receivers can be written as follows, if the channel is flat:

$$
\begin{bmatrix} y_1 \\ y_2 \\ \vdots \\ y_{N_r} \end{bmatrix} = \begin{bmatrix} h_{11} & h_{12} & \cdots & h_{1N_t} \\ h_{21} & h_{22} & \cdots & h_{2N_t} \\ \vdots & \vdots & \ddots & \vdots \\ h_{N_r1} & h_{N_r2} & \cdots & h_{N_rN_t} \end{bmatrix} \begin{bmatrix} x_1 \\ x_2 \\ \vdots \\ x_{N_t} \end{bmatrix}. \tag{4.12}
$$

Using a more compact matrix representation,

$$
\mathbf{y}_{N_r \times 1} = \mathbf{H}_{N_r \times N_t} \mathbf{x}_{N_t}.
$$

In the above we replace s with the transmitted signal vector $\mathbf{x}$ as the *space-time coded* version of the symbol sequence.

For broadband scenarios that involve L distinct multipath clusters,

$$
\mathbf{y}(t) = \sum_{l=1}^{L} \mathbf{H}_l \mathbf{x}(t - \tau_l) \tag{4.13}
$$

where each $\mathbf{H}_l$ is an $N_r \times N_t$ matrix.

As will become clear in the following discussion, the *capacity* of the MIMO channel depends on the characteristics of the channel, both statistically and algebraically. While a common assumption of the channel is that $\mathbf{H}$ is element-wise independent complex Gaussian random variables with zero mean, most practical channels do not fall under this category.

- Indoor micro MIMO: the ideal situation of MIMO involves rich scattering at both the transmitter side and the receiver side. In this case, it is reasonable to assume that each transmitter-receiver pair experiences independent fading and $\mathbf{H}$ is complex Gaussian with $\mathbf{R_H} = E\left\{\mathbf{HH}^H\right\} = \mathbf{I}$.

- Outdoor macro MIMO: Figure 4.5 illustrates an outdoor scenario where there are L multipaths from the base antenna array having distinct angles. Here the base-station is placed high above ground and is not surrounded by local scatters. In contrast, the user station is assumed to have uncorrelated fading at the antennas due to the fact that usually a large number of local scatters exists around the mobile antenna.Hence the outdoor downlink channel matrix has *uncorrelated rows and correlated columns*, whereas the uplink channel matrix has *uncorrelated columns and correlated rows*.

- Distributed MIMO: a few other systems can be modeled as MIMO channels. One example is the *relay network* where a source signal is delivered to destination(s) through multiple relay nodes [8]. Other examples include distributed sensor networks [9] and single-frequency broadcasting networks.

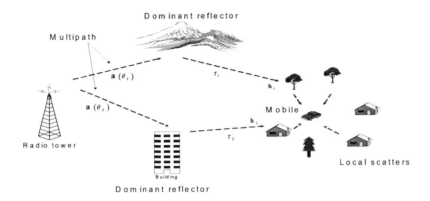

Figure 4.5: An outdoor wide-area MIMO scenario

A fundamental question in MIMO systems is its "capacity" with respect to the number of multiple transmitter antennas and receiver antennas. The answer can be found from an information theoretical viewpoint.

Given a MIMO channel, the mutual information between $\mathbf{x}$ and $\mathbf{y}$ is:

$$
\begin{aligned}
I(\mathbf{x}; \mathbf{y}) &= h(\mathbf{y}) - h(\mathbf{y}|\mathbf{x}) = h(\mathbf{y}) - h(\mathbf{n}) \\
&= h(\mathbf{y}) - \sum_{i=1}^{N_r} h(n_i) \\
&\leq \log((2\pi e)^{N_r} |\mathbf{R_y}|) - \sum_{i=1}^{N_r} h(n_i) \\
&= \log((2\pi e)^{N_r} |\mathbf{R_y}|) - \frac{1}{2} \log((2\pi e)^{N_r} |\sigma^2 \mathbf{I}_{N_r}|) \\
&= \log \frac{|\mathbf{R_y}|}{\sigma^2}
\end{aligned}
$$

where $\sigma^2 \mathbf{I}_{N_r} = E\{\mathbf{n}\mathbf{n}^H\}$, and $\mathbf{R_y} = E\{\mathbf{y}\mathbf{y}^H\} = \mathbf{H}\mathbf{R_x}\mathbf{H}^H + \sigma^2 \mathbf{I}_{N_r}$.

As a result, the MIMO capacity is given by

$$
C = \log \left(\left| \mathbf{I} + \frac{\mathbf{H}\mathbf{R_x}\mathbf{H}^H}{\sigma^2} \right| \right) \tag{4.14}
$$

If the channel characteristics $\mathbf{H}$ are known to the transmitter (i.e., *informed*), it is possible to manipulate the transmit signal variance $\mathbf{R_x}$ to maximize the channel capacity. On the other hand, if the channel information is not available at the transmitter (i.e., *uninformed*), we let $\mathbf{R_x} = \frac{P}{N_t}\mathbf{I}_{N_t}$, i.e., equal power allocation over N_t transmitters with *fixed* total transmit power P . Then

$$C = \log\left(\left|\mathbf{I_R} + \frac{P}{N_t}\frac{\mathbf{HH}^H}{\sigma^2}\right|\right)$$

Denote $\rho = \frac{P}{\sigma^2}$ as the transmit signal to noise power ratio, then the capacity with parameter N_r and N_t per unit bandwidth is

$$C(N_t, N_r) = \log\left(\left|\mathbf{I} + \rho\frac{\mathbf{HH}^H}{N_t}\right|\right). \tag{4.15}$$

- *SISO:* $\mathbf{H} = h$. With one transmit antenna and one receiver antenna, the channel capacity reduces to the classic result:

$$C(1, 1) = \log(1 + \rho|h|^2)$$

Observation: the capacity increases logarithmically with the SNR. Each extra bps/Hz requires roughly a doubling of the TX power. For example, to go from 1bps/Hz to 11bps/Hz, the TX power must be increased by roughly 1000 times!

- *MISO:* $\mathbf{H} = [h_1, h_2, ..., h_{N_t}]$. This scenario is typically seen in cellular downlink with an antenna array at the base-station and a single antenna at the user terminal.

$$C(N_t, 1) = \log\left(1 + \rho\frac{|\mathbf{H}|^2}{N_t}\right) = \log\left(1 + \rho\frac{\sum_{i=1}^{N}|h_i|^2}{N_t}\right)$$

Observation: with the total transmission power fixed, the capacity actually does not increase with additional transmission antennas (unless $\mathbf{H}$ is known at the transmitter, in which case transmit beamforming can be utilized to increase the SINR - see page 2). On the other hand, the *outage capacity* over a fading channel does improve due to the increased diversity.

- *SIMO:* $\mathbf{H} = [h_1, h_2, ...h_{N_r}]^T$. This scenario is typically seen in cellular uplink with an antenna array at the base-station and a single antenna at the user terminal.

$$C = \log\left(\left|\mathbf{I} + \rho\mathbf{HH}^H\right|\right).$$

Since $\left|\mathbf{I}_{N_r} + \rho\mathbf{HH}^H\right| = \left|\mathbf{I}_{N_t} + \rho\mathbf{H}^H\mathbf{H}\right|$, we get

$$C = \log\left(1 + \rho|\mathbf{H}|^2\right) = \log\left(1 + \rho\sum_{i=1}^{N_r}|h_i|^2\right).$$

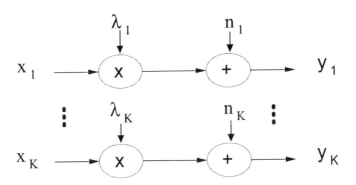

Figure 4.6: An equivalent parallel SISO representation of the MIMO channel.

Observation: the capacity increases logarithmically with respect to the number of antennas at the receiver side.

- *MIMO:* To gain insight to the MIMO scenario, let us perform an eigenvalue decomposition on the channel matrix as following

$$\mathbf{HH}^H = \mathbf{E\Lambda E}^H.$$

We obtain from (4.15)

$$C = \sum_{k=1}^{K} \log\left(\left|1 + \rho\frac{\lambda_k}{N}\right|\right).$$

where $\{\lambda_k\}$ are the K non-zero eigenvalues of the channel covariance. It is seen that the MIMO channel can be decomposed into K multiple parallel SISO channels that can be used to deliver information simultaneously, indicating the potentially linear increase in capacity with respect to $K \leq \min(N_r, N_t)$. An equivalent representation of the MIMO channel is given in Figure 4.6.

Note that the distribution of $\{\lambda_k\}$ is determined by the condition of the channel matrix. It can be shown that the maximum capacity is achieved when $\lambda_1 = \lambda_2 = \cdots = \lambda_k$, in which case

$$C = K \log\left(\left|1 + \frac{\rho}{N}\right|\right).$$

- Broadband MIMO: from (4.13), the channel response in frequency domain is given by.

$$\mathbf{H}(\omega) = \sum_{l=1}^{L} \mathbf{H}_l e^{-j\omega l}$$

The total capacity is obtained by replacing $\mathbf{H}$ in (4.14) with $\mathbf{H}(\omega)$, and integrating over the frequency band [7].

4.4   Space-time coding

To capture the potential of MIMO channels promised by (4.14) and (4.15), multiple-antenna transmission schemes as well as sophisticated MIMO receiver techniques need to be developed. In this section, we discuss the basic principles of space-time coding and lay some groundwork for Section 4.5.

In the most generic form, a space-time encoder provides a mapping between the information-bearing symbols $\{s\}$ to an N_t -dimensional stream $\mathbf{c}$ ($\mathbf{x} = \mathbf{c}$ if no additional operation is involved)

$$\{s\} \Longrightarrow \{\mathbf{c}\}$$

Each element of $\mathbf{x}$ is transmitted simultaneously from a different transmit antenna as specified in (4.12). We define the space-time coding rate R_{ST} as

$$R_{ST} = \text{the number of symbols/transmission}$$

The actual data rate (bit/second) depends on how many bits each symbol carries.

4.4.1   Spatial multiplexing

The spatial multiplexing scheme sends independent data streams over the individual TX channels. The best known example is the Bell Labs Layered Space-time Architecture (BLAST) which has the following mapping [3]:

$$\mathbf{c}(k) = \mathbf{x}(k) = \begin{bmatrix} s(N_t k + 1) \\ s(N_t k + 2) \\ \vdots \\ s(N_t k + N_t) \end{bmatrix}.$$

As a result, spatial multiplexing offers the maximum rate of $R_{ST} = N_t$. If $N_r \geq N_t$, the symbol streams can be recovered by inverting the MIMO channel:

$$\widehat{\mathbf{s}} = \mathbf{H}^\dagger \mathbf{y}$$

Spatial multiplexing is rather simple and has good behavior in environment with high scattering. The data rate can be extremely high in favorable channel conditions (high SNR and high scattering). On the other hand, its performance decreases with increasing spatial correlation in the channel. In addition, on the receiver side we need at least as many antennas as on the transmitter side

4.4.2 Orthogonal space-time block coding

Alamouti [4] discovered a remarkable space-time block coding scheme that achieves both SINR gain and diversity gain with two transmit antennas. The scheme was later generalized into a class of *orthogonal space-time block* codes (OSTBC) that promises full transmit diversity and simple reception [11].

The Alamouti code is a rate-1 $(R_{ST} = 1)$ orthogonal space-time block code with the following mapping

$$\mathbf{s} = \begin{bmatrix} s_1 \\ s_2 \end{bmatrix} \Longrightarrow \mathbf{X} = [\ \mathbf{x}(1) \quad \mathbf{x}(2) \ ] = \begin{bmatrix} s_1 & -s_2^* \\ s_2 & s_1^* \end{bmatrix}$$

The two symbols are transmitted over two time units from the two antennas. At the receiver side, the received signal over two consecutive symbols are

$$\begin{aligned} y(1) &= h_1 s_1 + h_2 s_2 + n_1 \\ y(2) &= -h_1 s_2^* + h_2 s_1^* + n_2 \end{aligned}$$

By stacking the two consecutive received signals as

$$\mathbf{r} = [\ y(1) \quad y^*(2) \ ]^T$$

we have

$$\mathbf{r} = \mathbf{Gs} + \mathbf{n}; \quad \mathbf{G} = \begin{bmatrix} h_1 & h_2 \\ h_2^* & -h_1^* \end{bmatrix}$$

where $\mathbf{n} = [\ n_1 \quad n_2 \ ]^T$.

The maximum likelihood decoder is given by

$$\hat{\mathbf{s}} = \arg \min_{\hat{\mathbf{s}} \in \mathcal{C}} ||\mathbf{r} - \mathbf{G}\hat{\mathbf{s}}||^2$$

Notice that the $\mathbf{G}$ matrix is orthogonal, i.e., $\mathbf{G}^H \mathbf{G} = \rho \mathbf{I}$, where $\rho = |h_1|^2 + |h_2|^2$. We may modify the signal vector and reduce the ML decoding to:

$$\begin{aligned} \hat{\mathbf{r}} &= \mathbf{G}^H \mathbf{r} = \rho \mathbf{s} + \hat{\mathbf{n}} \\ \hat{\mathbf{s}} &= \arg \min_{\mathbf{s} \in \mathcal{S}} ||\hat{\mathbf{r}} - \rho \mathbf{s}||^2 \end{aligned}$$

Surprisingly, the decoding rule in the above ML formulation reduces to two independent, much simpler decoding rules for s_1 and s_2. Because the symbols are transmitted from two antennas, it is easy to verify that an order-2 transmit diversity is achievable. On the other hand, since no channel information is

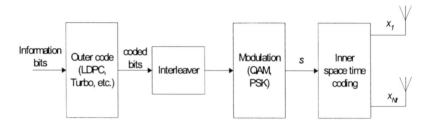

Figure 4.7: An MIMO system block diagram

utilized at the transmit, the SINR gain of Alamouti code should not be as significant as downlink beamforming (4.11). Such is indeed the case as the SNR of the decoder output is given by

$$SNR = \frac{(|h_1|^2 + |h_2|^2)\, \sigma_s^2}{\sigma_n^2} \frac{\sigma_s^2}{2} \Rightarrow E\{SNR\} = \frac{\sigma_s^2 \sigma_a^2}{\sigma_n^2} \qquad (4.16)$$

A factor of 2 is introduced in the SNR calculation to normalize the total transmitted power to σ_s^2. Compared to (4.11), it is obvious that the Alamouti scheme is at a 3dB SNR disadvantage to the beamforming scheme.

Orthogonal space-time block codes for $N_t > 2$ are available, but only for R_{ST} < 1 symbols/transmission [11]. In addition, the OSTC does not introduce any coding gain. Non-orthogonal space-time block codes based on linear complex field enable full diversity without any rate loss (1 symbol/transmission)

4.4.3   Concatenated ST transmitter

For actual systems, STC is used in combination with an outer channel encoder as depicted in Figure 4.7. Essentially, it forms a serial concatenation of turbo encoder where an iterative detection and decoding strategy can be employed at the receiver side [6].

- The channel encoder, e.g., the LDPC or convolutional encoders, provides an outer code that can be decoded by a soft-input soft-output decoder.

- The interleaver interleaves the coded bit stream on a block-by-block basis. The output stream is mapped onto QAM or M-ary PSK symbols.

- A linear space-time encoder provides the tradeoff between the data rate and diversity order, with extreme cases being the BLAST (highest rate) and the OSTBC (full diversity).

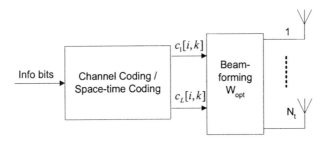

Figure 4.8: Beamforming with ST coding

A wide variety of decoding schemes can be used, offering tradeoffs between performance and complexity [6]. The overall rate of the transmitter is determined by

$$R_{TX} = R_{\text{outer code}} \times R_{\text{QAM}} \text{ [bits/symbol]} \times R_{ST} \text{ [symbol/s]}.$$

Determining the optimum combination for a given channel involves at least three parameters: the outer channel coding rate, the modulation scheme, and the STC rate. The tradeoffs between the three sub-modules in a space-time transmitter remain to be fully investigated.

4.4.4 Beamforming with ST coding

In some applications where MIMO channel information is known at the transmission side (even partially), transmit beamforming in addition to the space-time coding can be employed to further improve the system performance. A block diagram of the combination of space-time coding with transmit beamforming is illustrated in Figure 4.8. Transmit beamforming utilizes certain knowledge of the channel to achieve transmit antenna array gains. The optimal transmitter design has previously been investigated in [12][13] based on a capacity criterion. To understand the potential, let us investigate the change of capacity if the transmitter is informed.

From (4.14), we can see that by letting

$$\{s\} \Longrightarrow \{\mathbf{c}\} \Longrightarrow \mathbf{x} = \mathbf{Wc}$$

where $\mathbf{c}$ is the space-time coded vector which is assumed to be i.i.d. statistically. One can change the covariance matrix of $\mathbf{x}$ by adjusting the beamforming matrix $\mathbf{W}$. If the channel matrix $\mathbf{H}$ is fully known at the transmitter, the optimum beamformer that maximizes the channel capacity is then found to be

$$\mathbf{W}_{opt} = \arg_{\mathbf{W}} \max \left| \mathbf{I_R} + \frac{\mathbf{HWW}^H \mathbf{H}^H}{\sigma^2} \right|$$
$$\text{subject to} \qquad \text{tr}[\mathbf{WW}^H] = 1$$

In many applications, full knowledge of the channel is often not available. Rather, the statistical information of the channel may be estimated. For example, we can assume that the channel can be expressed as

$$\mathbf{H} = \mathbf{H}_r \mathbf{R_H}^{1/2} \qquad (4.17)$$

where $\mathbf{H}_r$ is i.i.d complex Gaussian, and $\mathbf{R_H}$ is the transmit antenna correlation matrix. In this case, statistical beamforming in conjunction with space-time coding can be applied. Sampath and Paulraj [5] and Zhou and Giannakis introduced the notion of eigen-beamforming using orthogonal space-time block codes and vector channel covariance matrices [14]. Focusing on symbol-by-symbol detection, the optimal beamforming was designed in [16] to minimize symbol error rate (SER).

Consider a size $L \times K$ space-time block code

$$\mathbf{C} = \begin{bmatrix} \mathbf{c}(1) & \mathbf{c}(2) & \mathbf{c}(K) \end{bmatrix}.$$

Instead of transmitting $\mathbf{C}$ directly through N_t antennas, an $N_t \times L$ beamforming transmit beamforming matrix $\mathbf{W}$ is applied, given the following transmitted signal

$$\mathbf{X} = \mathbf{WC}$$

Then the received signal becomes

$$\mathbf{Y} = \mathbf{HX} = \mathbf{HWC}$$

The optimum *statistical* beamformer is the one that optimizes the error performance. Based on pairwise error probability of the receiver with ML detection established in [10], it is straightforward to show that the probability of the decoder deciding in favor of code word $\mathbf{C}^e$, when in fact the code matrix $\mathbf{C}$ was transmitted, is upper bounded as

$$P_e \leq P(\mathbf{C} \to \mathbf{C}^e) \leq \exp\left(-\|\mathbf{HW}(\mathbf{C} - \mathbf{C}^e)\|_F^2 / 4N_o\right) \qquad (4.18)$$

Averaging across all possible channel realizations, the average PEP is shown as

$$\overline{P}_e = E_{\mathbf{H}}\{P_e\} \leq c \left| \mathbf{I} + \frac{(\mathbf{C} - \mathbf{C}^e)(\mathbf{C} - \mathbf{C}^e)^H \mathbf{W}^H \mathbf{R_H W}}{4N_o} \right|^{-N_r}$$

from which the optimum beamformer $\mathbf{W}_{opt}$ can be derived.

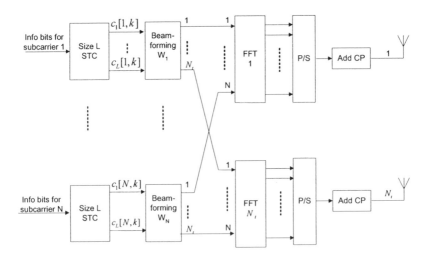

Figure 4.9: STC with beamforming in OFDM

4.4.5 ST beamforming in OFDM

The ST beamforming principles can be applied to broadband system using OFDM modem. By transforming the broadband channels into a set of parallel narrowband subchannels, the space-time coding beamforming can take the canonical form depicted in Figure 4.9.

4.5 Wide-area MIMO beamforming

Much of the work to date on STC-OFDM assumes a multipath channel in which fading from each transmit antenna to any receive antenna is independent. Such is valid only in a rich scattering environment. In many wide-area applications, the base station (BTS) is usually placed high above the ground and is not surrounded by scatters. The multipaths at the basestation antenna array will have distinct angles with each multipath corresponding to a distant scatter. As a result, the fadings in the MIMO channel are correlated with each other. Since the BTS-to-user transmission is more important in asymmetric applications, there is a strong need for an optimized space-time downlink modem for such channel conditions. In this section, we present a downlink MIMO beamforming scheme suitable for outdoor applications.

4.5.1 Data model

Negi *et al.* introduced a downlink beamforming STC scheme specifically for outdoor channels [15]. Here, we describe a low-complexity transmission scheme for broadband OFDM operations. The approach offers the benefits of both transmit beamforming and space-time coding to an OFDM system without minimum increase in complexity. In the outdoor channel model, the optimum beamformer is shown to be identical for all subcarriers in the OFDM system, allowing simple time domain implementation of the transmit beamformer and a simpler receiver.

Referring to Figure 4.5 which depicts the downlink scenario under consideration, the $N_r \times 1$ received signal at the mobile station can be expressed as [15]

$$\mathbf{y}(k) = \sum_{l=1}^{L} \mathbf{h}_l \mathbf{a}^H(\theta_l) \mathbf{x}(k - \tau_l) + \mathbf{v}(k)$$

where k denotes the discrete time index and τ_l is the discrete delay for path l; $\mathbf{h}_l = [h_{l1} \cdots h_{lN_r}]^T$ is the $N_r \times 1$ vector of fading channel gains associated with the lth multipath; $\mathbf{a}(\theta_l)$ is the $N_t \times 1$ vector representing the array response for the lth path at angle θ_l; $\mathbf{x}(k) = [x_1(k) \cdots x_{N_t}(k)]^T$ is the $N_t \times 1$ vector of the transmitted signal, and $\mathbf{v}(k)$ is the $N_r \times 1$ noise vector at the receive antennas.

In the above model, the fading gain vectors $\{\mathbf{h}_l\}$ for the L distinct paths are assumed to be uncorrelated. Also, their elements are assumed to have independent and identically distributed Rayleigh fading due to the abundant local scatters around the mobile. Each $\mathbf{h}_l \mathbf{a}^H(\theta_l)$ forms an $N_r \times N_t$ MIMO matrix $\mathbf{H}$, whose columns are correlated. We further assume that

- the BTS has knowledge of the multipath angles $\{\theta_l\}$ which are slow changing in an outdoor environment;

- the BTS does not know the fading gains $\mathbf{h}_l$ since they may be fast varying due to the motion of the terminal and environment variations;

- the mobile has full knowledge of channel state information (CSI) through channel estimation.

With an N-subcarrier OFDM, the input-output relation for the ith subcarrier in this system can be expressed as

$$\mathbf{y}(i, k) = \sum_{l=1}^{L} [\mathbf{h}_l \mathbf{a}^H(\theta_l) e^{-j\frac{2\pi}{N}\tau_l i}] \mathbf{x}(i, k) + \mathbf{v}(i, k), i = 1, \cdots, N$$

Written in matrix form, we have

$$\mathbf{y}(i, k) = \mathbf{H} \mathbf{D}_i \mathbf{A}^H \mathbf{x}(i, k) + \mathbf{v}(i, k)$$

where $\mathbf{H} = [\mathbf{h}1 \cdots \mathbf{h}_L]$, $\mathbf{A} = [\mathbf{a}(\theta_1) \cdots \mathbf{a}(\theta_L)]$, and

$$\mathbf{D}_i = diag\left\{ e^{-j\frac{2\pi}{N}\tau_1 i}, \cdots, e^{-j\frac{2\pi}{N}\tau_L i} \right\}.$$

For conventional STC-OFDM transmission over uncorrelated broadband MIMO channels, a size N_t space-time code is needed on each subcarrier to achieve the full spatial diversity. However in our channel model there are only L independent paths, the total order of diversity available from the channel is N_rL instead of N_rN_tL, suggesting a STC of size L is sufficient [15]. Without utilizing the channel structure or the angle information, the conventional STC scheme is sub-optimum in terms of both performance (i.e., no beamforming gain) and complexity (oversized STCs and FFT operations). In the following, we will develop a scheme to achieve better performance at lower complexity.

4.5.2 Uncoded OFDM design criterion

Since the maximum diversity in the system is N_rL, we start with a size L space time code which is able to capture L-order diversity in uncorrelated MIMO channels. Now for each subcarrier, we employ a size $L \times K$ space-time code $\mathbf{C}_i = \begin{bmatrix} \mathbf{c}_i(1) & \mathbf{c}_i(2) & \cdots & \mathbf{c}_i(K) \end{bmatrix}$. Combined with a transmit beamforming matrix $\mathbf{W}_i$, the transmitted signal on the ith subcarrier is

$$\mathbf{X}_i = \mathbf{W}_i\mathbf{C}_i$$

where $\mathbf{X}_i$ is the transmitted symbol on the ith subcarrier over K OFDM symbols. Then the received signal at subcarrier i is

$$\mathbf{Y}_i = \mathbf{HD}_i\mathbf{A}^H\mathbf{W}_i\mathbf{C}_i + \mathbf{V}_i$$

where $\mathbf{Y}_i = \begin{bmatrix} \mathbf{y}_i(1) & \mathbf{y}_i(2) & \cdots & \mathbf{y}_i(K) \end{bmatrix}$. To find the best transmission scheme in terms of error performance, let us first derive the performance criteria of the above STC coded system. From (4.18), the probability of the decoder deciding in favor of code word $\mathbf{C}_i^e$, when in fact the code matrix $\mathbf{C}_i$ was transmitted, is upper bounded as

$$P_e \leq P(\mathbf{C} \to \mathbf{C}^e) \leq \exp\left(-|\mathbf{HD}_i\mathbf{A}^H\mathbf{W}_i(\mathbf{C}-\mathbf{C}^e)|_F^2/4N_o\right) \qquad (4.19)$$

Averaging across all possible channel realizations, the average PEP is shown in Appendix I as

$$\overline{P}_e = \left(\frac{1}{4N_o}\right)^{-N_rL}\left|\mathbf{D}_i\mathbf{A}^H\mathbf{W}_i(\mathbf{C}-\mathbf{C}^e)(\mathbf{C}-\mathbf{C}^e)^H\mathbf{W}_i^H\mathbf{AD}_i^H\right|^{-N_r} \qquad (4.20)$$

Now we optimize $\mathbf{W}_i$ for each subcarrier to minimize error probability. Since our objective is to design the optimum beamforming matrix for any pre-chosen space time codes $\mathbf{C}$, we can assume $\left|(\mathbf{C}-\mathbf{C}^e)(\mathbf{C}-\mathbf{C}^e)^H\right|$ to be a constant which does not affect the optimization result. With this assumption, the optimization problem reduces to

$$\begin{array}{ll} \min & \left|\mathbf{D}_i\mathbf{A}^H\mathbf{W}_i\mathbf{W}_i^H\mathbf{AD}_i^H\right| \\ \text{subject to} & tr[\mathbf{W}_i\mathbf{W}_i^H] = L \end{array}$$

The constraint arises due to the total transmitted power over N_t antennas. Under unit transmit power assumption, a size L space-time code allocates an average power of $1/L$ to each of its L output symbols, i.e.,

$$E\left\{\mathbf{c}(k)\mathbf{c}^H(k)\right\} = \mathbf{I}/L$$

The optimum solution is given by the following proposition.

Proposition 3 *The optimum beamforming matrix* $\mathbf{W}_{opt}$ *for each subcarrier is identical, and is composed of the first L right singular vectors of matrix* $\mathbf{A}^H$.

Proof. Notice that matrix $\mathbf{D}_i$ is orthogonal; the maximization criterion can be simplified as

$$\min\left|\mathbf{D}_i\mathbf{A}^H\mathbf{W}_i\mathbf{W}_i^H\mathbf{A}\mathbf{D}_i^H\right| = \min\left|\mathbf{A}^H\mathbf{W}_i\mathbf{W}_i^H\mathbf{A}\mathbf{D}_i^H\mathbf{D}_i\right| = \min\left|\mathbf{A}^H\mathbf{W}_i\mathbf{W}_i^H\mathbf{A}\right|$$

It is obvious that the optimum solution is only decided by the matrix $\mathbf{A}$ and is independent of subcarrier index i. Following the similar derivation as in [15], we can show the optimum solution is to choose $\mathbf{W} = \mathbf{Q}_+\mathbf{\Phi}$, where $\mathbf{\Phi}$ is an arbitrary orthogonal matrix, and $\mathbf{Q}_+$ is comprised of the first L right eigenvectors of matrix $\mathbf{A}^H$. A natural choice is to let $\mathbf{\Phi}$ be an identity matrix, which leads to $\mathbf{W} = \mathbf{Q}_+$. ∎

Due to the fact that beamforming vectors are identical for all subcarriers, the computation complexity for finding the beamforming matrix is dramatically reduced. In fact, beamforming can be performed in the time domain in an OFDM system, which can further reduce the transmitter complexity.

Proposition 4 *The outdoor STC-beamforming scheme can be realized with L IFFT operations, following by a size $L \times N_t$ beamformer as shown in Figure 4.10.*

Fig. 4.10 illustrates the TX structure of a STC-beamforming OFDM system. Note when the number of antennas N_t is greater than the number of path L, this transmission scheme can reduce the transmitter complexity, requiring only L FFT computations instead of N_t FFTs as in the conventional STC-OFDM system.

With the optimum beamforming transmission scheme, the effective channel at the receiver for the ith subcarrier in an OFDM system becomes the standard uncorrelated MIMO channel,

$$\mathbf{Y}_i = \mathbf{H}_{i,eff}\mathbf{C}_i + \mathbf{V}_i \tag{4.21}$$
$$\mathbf{H}_{i,eff} = [\sigma_1\mathbf{h}_1' \cdots \sigma_L\mathbf{h}_L'] \tag{4.22}$$

where σ_l are the L singular values of the matrix $\mathbf{A}^H$, and $[\sigma_1\mathbf{h}_1' \cdots \sigma_L\mathbf{h}_L'] = \mathbf{H}\mathbf{D}_i\mathbf{U}$, where $\mathbf{U}$ is the left singular vectors of $\mathbf{A}^H$. Since both $\mathbf{D}_i$ and $\mathbf{U}$

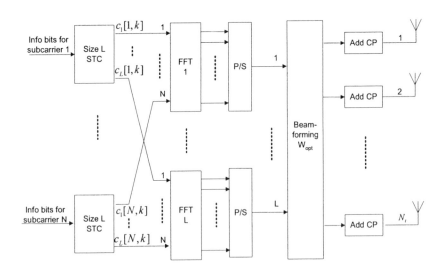

Figure 4.10: A simplified STC with beamforming architecture

are orthogonal matrices and $\mathbf{H}$ has i.i.d complex Gaussian elements, $\{\mathbf{h}'_l\}$ are uncorrelated Rayleigh fading random vectors. Therefore, it is straightforward to show that with an L-diversity space-time code, a diversity order of $N_r L$ is achieved.

To perform coherent detection, the receiver needs to estimate each subcarrier's effective channel $\mathbf{H}_{i,eff}$. With estimated channel coefficients, standard ML decoding can be employed to detect $\mathbf{C}_i$ from $\mathbf{Y}_i$ optimally,

$$\widehat{\mathbf{C}}_i = \arg \min_{\mathbf{C}_i} |\mathbf{Y}_i - \mathbf{H}_{i,eff} \mathbf{C}_i| \, .$$

Note the complexity of ML decoding is exponentially proportional to the size of the STC, which is the number of available paths L. When L is less than the available antennas N_t at the basestation, a reduction in decoding complexity compared to the conventional STC-OFDM of the order of $N_t - L$ is achieved.

In addition to lower encoder and decoder complexity as explained above, the outdoor STC-beamforming scheme also has better error performance than the conventional STC-only scheme. Since the two schemes use different size space-time codes, it is difficult to compare the performance in terms of pairwise error probability in (4.20), which depends on the specific space-time code design. Instead, we use the average receiving SNR as a metric to compare the two schemes. The following proposition states that the STC-beamforming scheme enjoys an SNR gain of $10 \log \left(\frac{N_t}{L} \right)$ dB.

Proposition 5 *For a system with N_t transmit antennas and L distinct paths $(N_t > L)$, the proposed STC-beamforming scheme has a receiving SNR gain of $10 \log \left(\frac{N_t}{L} \right) dB$ compared with the standard STC only scheme.*

Proof. See Appendix II.   ∎

4.5.3   Coded OFDM design criterion

In the previous section, we have demonstrated that combining beamforming with space-time codes can provide full diversity, achieve an SNR increase, and at the same time reduce the transmitter and receiver complexity. As mentioned in Section 4.4.3, an outer channel code is often employed in practical systems to combat frequency selectivity. A natural question here is how the outer channel encoder affects the ST-beamforming design in the outdoor ST-beamforming scheme.

Figure 4.11 shows the coded OFDM system under consideration. The information bits are first encoded with an outer channel code. We assume that the coding block spans more than one OFDM symbol and that the coded bits are fully interleaved before they are mapped to constellation points. A size $\widehat{L}$ space time code is employed on each subcarrier followed by standard OFDM modulation. As shown previously, beamforming can be performed in the time domain, which reduces the number of FFT operations to $\widehat{L}$. In this section, we investigate the choice of $\widehat{L}$ and the beamformer **W** in the presence of an outer channel encoder.

Paths with distinct delay

We first consider the case when all multipaths have distinct delays. The following proposition states that with a strong outer channel coding, all available diversity can be captured without a space-time code.

Proposition 6 *In a coded OFDM system, full diversity can be achieved by coding across subcarriers when all multipaths have distinct delays. Consequently no space-time code is needed and the beamforming vector should be chosen as the first right singular vector of the matrix $\mathbf{A}^H$.*

Proof. See Appendix III.   ∎

Intuitively, the above result can be understood from the fact that the composite channels across subcarriers contain different linear combinations of uncorrelated paths. Consequently, the outer channel code can capture all angle diversity in the frequency domain. At the same time, by transmitting all energy along the first right singular vector, the highest SNR gain is attained.

Paths with same delay

Proposition 6 shows that when all paths in the outdoor model have different arrival time, full diversity can be captured by using an outer code across subcarriers. In other words, the angle diversity is readily transferred into the frequency diversity in the OFDM systems. Combined with a proper beamformer, the transmission scheme can achieve both diversity and SNR gain as well. However, such a scheme can not fully exploit the available diversity in the channel when several paths have the same delay. This loss of diversity can be shown by observing (4.21). Notice that when two paths, say $\mathbf{h}_1$ and $\mathbf{h}_2$, have the same delay τ_1 and τ_2, these two coherent paths will always be combined in the same way on all subcarriers. Therefore, the available diversity in the effective channel seen by the channel coding is reduced. To capture this otherwise lost diversity, a space-time code is needed. The size of the STC needed is dependent on the number of paths having the same delay, as shown in the following proposition.

Proposition 7 *In a coded OFDM system, a size $\widehat{L}$ STC is needed to capture the $\widehat{L}$ order spatial diversity when $\widehat{L}$ paths have same delays. Combined with outer channel coding across subcarriers and beamforming with the first $\widehat{L}$ right singular vectors, the transmission scheme is able to achieve full diversity and the largest SNR gain.*

Proof. Suppose $\widehat{L}$ paths have the same delay τ_1. Substituting the $N_t \times \widehat{L}$ beamformer matrix $\mathbf{W}$ into the channel model, we have the effective $N_t \times \widehat{L}$ channel matrix for the ith subcarrier as

$$\mathbf{H}_{i,eff} = \left[\widehat{\mathbf{h}}_{i,1} \cdots \widehat{\mathbf{h}}_{i,\widehat{L}} \right] = [\mathbf{h}_{i,1} \cdots \mathbf{h}_{i,L}] \, \mathbf{D}_i \, [\sigma_1 \mathbf{u}_1 \cdots \sigma_{\widehat{L}} \mathbf{u}_{\widehat{L}}]$$

It is easy to show that the $\widehat{L}$ columns of the effective channel matrix $\mathbf{H}_{i,eff}$ are uncorrelated by verifying that $E\left\{ \widehat{\mathbf{h}}_{i,j_1} \widehat{\mathbf{h}}_{i,j_2}^H \right\} = 0$, for $j_1 \neq j_2$. Therefore a diversity order of $\widehat{L}$ is obtained with a conventional size $\widehat{L}$ space-time code [10]. The remaining diversity from paths with distinct delays can be obtained by outer channel coding across subcarriers, as demonstrated in Proposition 6. ∎

Example 9 *Consider a 256-point OFDM system with a uniform linear array with $N_t = 4$ antennas at the transmitter. Figure 4.12 compares the performance of the STC-beamforming scheme, the conventional STC (STC-only), and the beamforming-only scheme. The number of paths $L = 2$. The angles of departure (AOD) and the delays of these two paths are $\{10°, 30°\}$ and $\{1, 10\}$ (samples), respectively. The space-time code employed is the orthogonal space-time block code. For the STC-beamforming scheme, a size-2 orthogonal STBC, i.e., Alamouti code, is combined with beamforming as described. For the beamforming only*

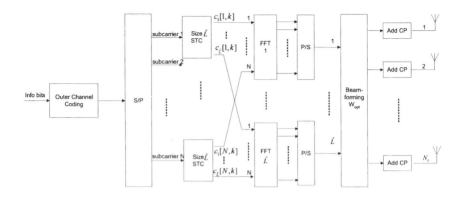

Figure 4.11: Concatinated coding in OFDM with beamforming

scheme, the beamformer vector is chosen as the first right singular vector, i.e. the dominant eigen-mode transmission. For the conventional STC transmission, a size-4 orthogonal STBC [11] with a rate of 1/2 is combined with 16QAM modulation to provide a spectral efficiency of 2 bits/symbol. The other two schemes use QPSK. This figure clearly shows that the STC-beamforming scheme has the best performance among all three. It has the same diversity order as the conventional STC scheme, while enjoying a SNR gain of about 5 dB. The larger SNR gain is partly due to the higher order modulation scheme used in the STC only scheme. Compared with the beamforming only scheme, the STC-beamforming scheme achieves a higher order of diversity as indicated by a steeper BER curve. It is also interesting to notice that BER curves for the conventional STC scheme and the beamforming only scheme cross over at SNR = 19dB. This is understandable since in the low SNR range, the beamforming-only scheme has better BER performance due to its SNR gain; while in the high SNR range, diversity becomes more important to performance. Although the size-4 STTC has a larger coding gain and a much higher decoding complexity, it is still outperformed by the 2-STTC combined with beamforming scheme. The SNR gain obtained by the STC-beamforming scheme is about 3dB, which is consistent with Proposition 5.

Example 10 *Performance of STC-beamforming transmission in coded OFDM systems are evaluated in Fig. 4.13(a) and (b). A 1/2 rate 64-state convolutional code is used to encode information bits across 256 subcarriers. Fig. 4.13(a) compares the performance of the STC-beamforming and the beamforming-only schemes for the case where the two paths in the channel have distinct delays $\tau_1 = 1$ and $\tau_1 = 10$. The space-time code used is the Alamouti code. As expected, the two schemes have the same diversity order while the STC-beamforming scheme slightly outperforms the other scheme by about 1dB. The reason for this gain is that the space-time code produces a flatter channel across all subcarriers, which*

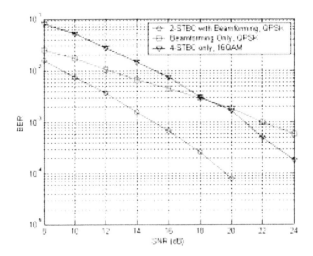

Figure 4.12: Performance of STC with beamforming vs. STC only schemes.

is more beneficial to the outer channel coding. Also plotted in the figure are the BER curves of the two schemes when there are $L = 4$ paths. Both schemes are able to pick up the higher order of diversity in this setup and provide similar performance. When several paths in the channel have same delays, the space-time code is needed to obtain all the diversity inherent to the channel. This effect is clearly demonstrated in the Fig. 4.13(b), where two channel setups are considered. For the first setup, there are 2 paths with the same propagation delay $\tau_1 = 1$. The number of antennas at the basestation is set to be 4. It is obvious that the beamforming only scheme can not achieve the diversity order of 2, even with the help of the FEC. Hence, it suffers a large performance loss in the high-SNR region. For the second channel setup, the number of distinct paths in the physical channel is assumed to be 4. The delays for these 4 paths are $\{1, 1, 10, 10\}$ respectively. Once again, it is seen that the Alamouti code combined with beamforming obtains the full diversity, while the beamforming-only scheme can only provide diversity of order 2. This is because the outer channel code across subcarriers can only exploit the diversity associated with paths having distinct delays, which in this case is 2.

4.6 Summary

Space-time processing adds a new dimension to OFDM that can dramatically enhance some key operational parameters in a wireless network. In this chap-

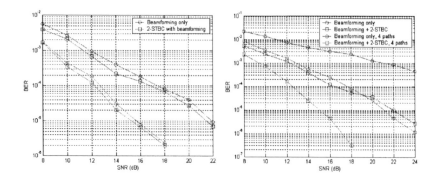

Figure 4.13: Beamforming with outer channel coding

ter, we discuss the basics of beamforming, space-time coding, the capacity of a MIMO system, and the use of STC/beamforming in OFDM systems. The activities in this area have led to many exciting breakthroughs, some of which have already been adopted into new wireless standards such as the WiFi and the WiMAX.

Appendix I: derivation of $\overline{P}_e$

Let $\mathcal{C} = \mathbf{A}^H \mathbf{W} (\mathbf{C} - \mathbf{C}^e)$, $d^2 = \|\mathbf{H}\mathcal{C}\|_F^2$, (4.19) then can be rewritten as

$$P_e \leq \exp\left(-d^2/4N_o\right)$$

Note d^2 can be expressed as

$$d^2 = |\mathbf{H}\mathcal{C}|_F^2 = \sum_{i=1}^{m} \mathbf{h}_i^H \mathcal{C}\mathcal{C}^H \mathbf{h}_i$$

where $\mathbf{h}_i$ is the ith column of $\mathbf{H}$. Substitute $\mathcal{C}\mathcal{C}^H$ with its eigenvalue decomposition $\mathbf{U}\mathbf{\Lambda}\mathbf{U}^H$, we have

$$d^2 = \sum_{i=1}^{N_r} \mathbf{h}_i^H \mathbf{U}\mathbf{\Lambda}\mathbf{U}^H \mathbf{h}_i = \sum_{i=1}^{N_r} \mathbf{g}_i^H \mathbf{\Lambda}\mathbf{g}_i = \sum_{i=1}^{N_r}\sum_{j=1}^{L} \lambda_j |g_{ij}|^2$$

Since $\mathbf{U}$ is an orthogonal matrix and $\mathbf{h}_i^H$ is a vector with i.i.d complex Gaussian elements, $\mathbf{g}_i^H = \mathbf{h}_i^H \mathbf{U}$ also has i.i.d complex Gaussian elements. Therefore, d^2 is the sum of $N_r \times L$ independent χ^2 random variables each with 2 degrees of freedom, and has the characteristic function as follows:

$$\psi_{d^2}(j\omega) = \left(\prod_{j=1}^{L} \frac{1}{1 - j\omega\lambda_j} \right)^{N_r}$$

Hence the average PEP is

$$
\begin{aligned}
\overline{P_e} &= \psi_{d^2}(j\omega)|_{\omega = \frac{-j}{4N_o}} = \left(\prod_{j=1}^{L} \frac{1}{1 + \frac{\lambda_j}{4N_o}} \right)^{N_r} \\
&\leq \left(\frac{1}{4N_o} \right)^{-N_r L} \left(\prod_{j=1}^{L} \lambda_j \right)^{-N_r} \\
&= \left(\frac{1}{4N_o} \right)^{-N_r L} |\mathcal{CC}^H|^{-N_r} \\
&= \left(\frac{1}{4N_o} \right)^{-N_r L} |\mathbf{A}^H \mathbf{W}(\mathbf{C} - \mathbf{C}^e)(\mathbf{C} - \mathbf{C}^e)^H \mathbf{W}^H \mathbf{A}|^{-N_r}
\end{aligned}
$$

Appendix II: proof of proposition 5

For the STC only scheme, each codeword matrix has dimension $N_t \times N_t$. The average receiving SNR on the ith subcarrier is

$$\overline{SNR}_{STC} = \frac{1}{N_t} \frac{1}{N_r N_o} E \left\{ tr \left[\mathbf{HD}_i \mathbf{A}^H \widetilde{\mathbf{X}} \widetilde{\mathbf{X}}^H \mathbf{AD}_i^H \mathbf{H}^H \right] \right\}$$

where the factor $\frac{1}{N_t}$ is due to the fact that each codeword spans N_t time slots, and $\frac{1}{N_r N_o}$ is the total noise power on N_r receive antennas. With the assumption that $E \left\{ \widetilde{\mathbf{X}} \widetilde{\mathbf{X}}^H \right\} = \mathbf{I}_{N_t}$,

$$
\begin{aligned}
\overline{SNR}_{STC} &= \frac{1}{N_t} \frac{1}{N_r N_o} E \left\{ tr \left[\mathbf{HD}_i \mathbf{A}^H \mathbf{AD}_i^H \mathbf{H}^H \right] \right\} && (4.23) \\
&= \frac{1}{N_t} \frac{1}{N_r N_o} E \left\{ tr \left[\mathbf{HD}_i \mathbf{U} \mathbf{\Sigma}_+^2 \mathbf{U}^H \mathbf{D}_i^H \mathbf{H}^H \right] \right\} \\
&= \frac{1}{N_t} \frac{1}{N_r N_o} \sum_{l=1}^{L} \sigma_i^2 E \left\{ tr[\mathbf{h}_l'^H \mathbf{h}_l'] \right\} \\
&= \frac{1}{N_r N_o} (\sigma_1^2 + \cdots + \sigma_L^2)
\end{aligned}
$$

where in the second step, we substitute $\mathbf{A}^H$ with its SVD, $\mathbf{A}^H = \mathbf{U}[\mathbf{\Sigma}_+|0]\mathbf{V}^H$; and in the third and the last steps, we use the facts that $[\mathbf{h}_1' \cdots \mathbf{h}_L'] = \mathbf{HD}_i \mathbf{U}$ has i.i.d complex Gaussian elements and $E \left\{ tr[\mathbf{h}_l'^H \mathbf{h}_l'] \right\} = N_r$.

For the STC-Beamforming scheme, the average receiving SNR is

$$\overline{SNR}_{STC-BF} = \frac{1}{L}\frac{1}{N_r N_o} E\left\{ tr\left[\mathbf{HD}_i \mathbf{A}^H \mathbf{W}_i \mathbf{C}_i \mathbf{C}_i^H \mathbf{W}_i^H \mathbf{AD}_i^H \mathbf{H}^H\right]\right\}$$

with a similar derivation as for the STC only scheme, it can be shown that

$$\overline{SNR}_{STC-BF} = \frac{1}{LN_o}(\sigma_1^2 + \cdots + \sigma_L^2) \tag{4.24}$$

Comparing (4.23) with (4.24), we have

$$10\log_{10}\left(\frac{\overline{SNR}_{STC-BF}}{\overline{SNR}_{STC}}\right) = 10\log_{10}\left(\frac{N_t}{L}\right).$$

Appendix III: proof of proposition 6

Transmitting along the first right singular vector of the matrix $\mathbf{A}^H$, i.e., the dominant eigenmode transmission, is optimum in terms of receiving SNR. This is because all transmit energy is concentrated along the direction corresponding to the largest singular value (hence the largest gain). All we need to show is that with this beamformer, the outer channel coding across subcarriers can catch all the available diversity.

To show this, consider the composite channel seen by the outer channel coding. Substituting the beamformer matrix $\mathbf{w} = [\mathbf{w}_1]$ into the channel model, we have the composite channel for the ith subcarrier as

$$\begin{aligned}
\widehat{\mathbf{h}}_i &= \sigma_1 \left[\mathbf{h}_1 \cdots \mathbf{h}_L\right] \mathbf{D}_i \mathbf{u}_1 \\
&= \sigma_1 \left[\mathbf{h}_1 \cdots \mathbf{h}_L\right] \begin{bmatrix} e^{-j\frac{2\pi}{N}\tau_1 i} & & \\ & \ddots & \\ & & e^{-j\frac{2\pi}{N}\tau_L i} \end{bmatrix} \begin{bmatrix} u_{11} \\ \vdots \\ u_{1L} \end{bmatrix}
\end{aligned}$$

where $\mathbf{u}_1$ and σ_1 are the first left singular vector and the largest singular value of matrix $\mathbf{A}^H$, respectively. Thus, the composite channel for the ith subcarrier is the sum of L uncorrelated Rayleigh fadings with different phase rotations. With the assumption that all L paths have different τ_l, these phase rotations are different across subcarriers.

Denote $d_{\min}$ the minimum Hamming distance of the outer FEC code. We further assume that $d_{\min}$ differences in coded bits result in $d_{\min}$ different symbols after modulation mapping, which is almost the case when a bit interleaver exists between the encoder and the modulator. Following the similar PEP derivation as in Appendix I, the average PEP for the coded system can be shown as

$$\overline{P}_e = \left(\prod_{j=1}^{r} \frac{1}{1 + \frac{\lambda_r}{4N_o}} \right)^{N_r}$$

where $\lambda_1, \lambda_2, \cdots, \lambda_r$ are the r nonzero eigenvalues of the correlation matrix $\mathbf{R}_{\widehat{h}_s} = E\left\{\widehat{\mathbf{h}}_s \widehat{\mathbf{h}}_s^H\right\}$, and $\widehat{\mathbf{h}}_s$ is the composite channel vector at subcarriers $\mathcal{S} = \{s_1, s_2, \cdots, s_{d_{min}}\}$, where two FEC codewords differ. Clearly, the rank of the correlation matrix $\mathbf{R}_{\widehat{h}_s}$, r, determines the diversity order that the scheme can achieve.

The rank r can be determined by plugging (4.21) into the correlation matrix,

$$\mathbf{R}_{\widehat{h}_s} = E\left\{\widehat{\mathbf{h}}_s \widehat{\mathbf{h}}_s^H\right\} = E\left\{\mathbf{B}_1 \left[\mathbf{h}_1 \cdots \mathbf{h}_L\right]_1^H \left[\mathbf{h}_1 \cdots \mathbf{h}_L\right] \mathbf{B}^H\right\} = \mathbf{B}\mathbf{B}^H$$

where

$$\mathbf{B} = \left[\mathbf{D}_{s_1}\mathbf{u}_1 \quad \mathbf{D}_{s_2}\mathbf{u}_1 \quad \cdots \quad \mathbf{D}_{s_{d_{min}}}\mathbf{u}_1 \right]$$

$$= \begin{bmatrix} u_{11} & & \\ & \ddots & \\ & & u_{1L} \end{bmatrix} \begin{bmatrix} e^{-j\frac{2\pi}{N}\tau_1 s_1} & \cdots & e^{-j\frac{2\pi}{N}\tau_1 s_{d_{min}}} \\ \vdots & \ddots & \vdots \\ e^{-j\frac{2\pi}{N}\tau_L s_1} & \cdots & e^{-j\frac{2\pi}{N}\tau_L s_{d_{min}}} \end{bmatrix}$$

Note that the second matrix in the above equation is nothing but a submatrix of an $N \times N$ DFT matrix. When the delay spread is uniform, it can be easily shown that the matrix is rank L, as any of its L columns form a Vandermonde matrix. In other cases, any L columns are shown to be almost always linearly independent. Hence the matrix is also rank L. Therefore, matrix $\mathbf{B}$, and consequently the correlation matrix $\mathbf{R}_{\widehat{h}_s}$ is rank L as long as d_{min} is larger than the number of the available paths L. This can be easily accomplished with an adequate FEC code.

Bibliography

[1] P. Tan and N. C. Beaulieu, "Reduced ICI in OFDM systems using the better than raised-cosine pulse," *IEEE Commun. Lett.*, vol. 8, no. 3, March 2004, pp 135-137.

[2] A. Papoulis, *Probability, random variables, and stochastic processes*, McGraw-Hill Inc, New York.

[3] J. G. Foschini, "Layered space-time architecture for wireless communication in fading environment when useing multi element antennas," *Bell Labs Tech. J.*, vol. 2, Autumn 1996, pp41-59.

[4] S. M. Alamouti, "A simple transmitter diversity scheme for wireless communications," *IEEE J. Select. Areas Commun.*, vol. 1 pp. 1451-1458, October 1998.

[5] H. Sampath and A. Paulraj, "Linear precoding for space-time coded systems with known fading correlations," *IEEE Comm. Letters*, vol 6, no. 6 pp 239-241, June 2002.

[6] S. Haykin, M. Sellathurai, Y. de Jong, and T. Willink, "Turbo-MIMO for Wireless Communications," *IEEE Communication Magazine*, pp 48 - 53, October 2004.

[7] E. G. Larsson and P. Stoica, *Space-time block coding for wireless communications*, Cambridge University Press, 2003.

[8] Y. Chang and Y. Hua, "Application of space-time linear block codes to parallel wireless relays in mobile ad hoc networks," *Proc of Asilomar Annual Conference on Signals Systems and Computers*, Pacific Grove, CA, Nov 9-12, 2003.

[9] J. N. Laneman and G. W. Wornell, "Distributed space-time-coded protocols for exploiting cooperative diversity in wireless networks," *IEEE Trans. Information Theory*, vol. 49, pp 2415-2425, Oct. 2003.

[10] V. Tarokh, N. Seshadri and A. R. Calderbank, "Space-time codes for high data rate wireless communication: performance criterion and code construction," *IEEE Trans. Information Theory*, vol 44, pp. 744-765, March 1998.

[11] V. Tarokh, H. Jafarkhani and A. R. Calderbank, "Space-time block codes from orthogonal designs," *IEEE Trans. Information Theory*, vol 45, no. 5, pp. 1456-1467, July 1998.

[12] S. A. Jafar, S. Vishwanath, and A. Goldsmith, "Channel capacity and beamforming for multiple transmit and receive antenna with covariance feedback," *Proc IEEE ISIT-2001*, p. 321, Washington, DC, June 2001.

[13] E. Visotsky and U. Madhow, "Space-time transmit precoding with imperfect feedback," *IEEE Trans Information Theory*, vol. 47, no. 6, pp 2632-2639, Sept. 2001.

[14] G. Jongren, M. Skoglund, and B. Ottersten, "Combining beamforming and orthogonal space-time block coding," *IEEE Trans. Inform. Theory*, vol.48, no.3, pp 611-627, March 2002.

[15] R.Negi, A. M. Tehrani and J. M. Cioffi, "Adaptive antennas for space-time codes in outdoor channels," *IEEE Trans. Communications*, vol 50, no. 12, pp 1918-1925, December 2002.

[16] S. Zhou and G. B. Giannakis, "Optimal transmitter eigen-beamforming and space-time block coding based on channel correlations," *IEEE Trans. Information Theory*, vol. 49, no. 7, pp 1673-1689, July 2003.

Chapter 5

Multiple Access Control Protocols

5.1 Introduction

The previous chapters primarily address the link-level issues between a transmitter and a receiver. In a multiuser environment, the radio resource must be shared among multiple users. In addition, to support bandwidth-demanding multimedia traffics, a wireless network must be able to serve a diverse set of users in highly dynamic, resource constrained and data intensive environments.

This chapter discusses the multiple access control/medium access control (MAC) related issues in OFDM-based broadband wireless networks. In a large scale wireless system, users' channel characteristics, mutual interference patterns, as well as traffic requirements are largely diverse. Therefore, each radio resource unit in the time-frequency-space dimension is likely to bear high utility value to certain users. Consequently, the total channel capacity or spectral efficiency of the network can be significantly increased through judicious resource allocation in MAC layer. This motivates the design of highly adaptive MAC layer protocols and algorithms that can cope with the channel and traffic dynamics.

5.2 Basic MAC protocols

There are two basic types of MAC protocols that are commonly used in modern wireless communication systems: contention based and non-contention based MAC protocols.

5.2.1 Contention based protocols

In a contention based MAC protocol, each terminal transmits in a decentralized way. There is no central controller, e.g., a base station, in the system that regulates when and on which channel the terminal should transmit. As a result, more than one terminals may transmit simultaneously, leading to signal collisions. Once collision occurs, the involved terminals back-off their transmissions, i.e., wait for a randomly selected time and retransmit. The two commonly used contention based protocols in wireless communications are ALOHA and carrier-sensing multiple access (CSMA). The difference between the two resides mostly in whether the protocol checks the availability of the wireless medium before the terminal starts transmission. In ALOHA, each terminal starts transmitting regardless of whether other terminals are using the same channel. While in CSMA, the terminal first senses the channel before transmission. If the channel is free, i.e., no presence of other terminals' signals, then it starts to transmit; otherwise, it will hold its transmission until the medium is free. Different variations of the two protocols exists. For example, the slotted ALOHA protocol allows the terminal to start transmission only at the beginning of the divided time slots. The CSMA/CA protocol exchanges short messages between the transmitter and the intended receiver before data communications. It also indicates the medium reservation period during the data transmission. These short messages and reservation broadcasting allow all the neighboring terminals (including the so-called "hidden nodes", please see the book Appendix on IEEE 802.11 for more details) to realize the busy medium and hold their transmissions until the existing communication finishes.

Both ALOHA and CSMA protocols have been adopted in many wireless systems. For example, GSM [1] uses the slotted ALOHA in the terminal's initial access process. When a terminal tries to establish a call or a connection with the base station, it first uses ALOHA protocol to send the access request to the base station. CSMA/CA is the key MAC scheme in wireless LAN networks [2]. In a wireless LAN network, all terminals sense the channels and exchange short messages before data transmissions.

Example 11 *IEEE 802.11 uses CSMA/CA based contention access scheme. The CSMA/CA is realized by a distributed coordination function (DCF). In Figure 5.1, if station A has a contention window of 15, and it randomly selects a value between 0 and 15, e.g., 4, then the station must wait for additional 4 time slots before it may transmit. During this period, the station keeps sensing the medium. Meanwhile, station B selects a shorter backoff value, e.g., 2 time slots, and starts transmission before A. Then A shall hold its data and update the medium status after it detects frames on the medium. Once B completes its transmission, A resumes its count down procedure and starts transmission when the remaining 2 time slots elapse. Interested readers are referred to the book Appendix on IEEE 802.11 or the IEEE web site for more details [2].*

Note that although the terminals using ALOHA or CSMA randomly pick their back-off time or sense the channel before transmissions, collisions may still occur. For instance, the collision of CSMA may happen when terminal A senses the medium as free while terminal B is transmitting. This may be caused by the propagation delay from terminal B to terminal A–A does not detect B's signal

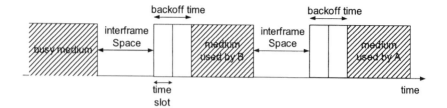

Figure 5.1: Medium contention of two users in a 802.11 network

because it has not yet reached terminal A. Generally speaking, the probability that multiple terminals having data to send at the same time increases as the terminal number increases, leading to a high collision probability. This has motivated the deployment of non-contention based MAC protocols in wireless systems.

5.2.2   Non-contention based MAC protocols

In non-contention based MAC protocols, a logic controller is needed to coordinate the transmissions of all the terminals. The controller informs each device when and on which channel it can transmit so that collisions can be avoided entirely. Non-contention based protocols can be further classified into two categories depending on whether the terminals are transmitting simultaneously using different channels (channelization) or they are transmitting sequentially using the same channel (non-channelization).

Non-channelization protocols

The commonly used non-channelization protocol in wireless system is polling protocol. Multiple access is coordinated by a master terminal which serves as the controller. The master terminal polls each terminal one by one to allow their data to transmit alternately. The role of the master terminal may be served in a round robin fashion by all the devices in the formed network.

Example 12 *IEEE 802.11 standard also has a polling based medium access method in addition to contention based method. The polling method is operated by a point coordination function (PCF). The access point polls each station and allows them to transmit alternately. The contention period and contention free period (polling period) operate alternately and the length ratio of these two periods can be preset by the network administrator.*

Example 13 *IEEE 802.15, also referred to as wireless personal area network (WPAN), employs a polling based medium access method. A device that coordinates the network (so-called piconet) is referred to as "master node", while others are called "slave nodes". The master node polls slave nodes alternatively*

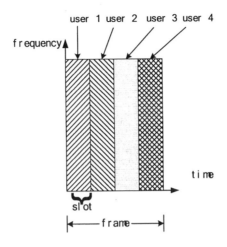

Figure 5.2: TDMA channelization

to allow them to transmit as permitted. All devices are synchronized with the master node's frequency hopping channels, using the clock of the master node in the piconet. A WPAN network addresses the short range communication ranging from 1 to 10m over a set of devices. Interested readers are referred to the standard body for more details [3].

Channelization protocols

Channelization protocols are the most commonly used protocols in cellular systems. The terminals transmit on the designated channels assigned to them. The channel is defined by how the frequency and time domains are divided and shared by the terminals. The channelization scheme determines the MAC scheme type. So far, three types of MAC schemes have been commercially used in the cellular world: time division multiple access (TDMA), code division multiple access (CDMA) and frequency division multiple access (FDMA). Figure 5.2 (TDMA), Figure 5.3 (CDMA) and Figure 5.4 (FDMA) illustrate these schemes. In the figures, different colors represent the occupancy of the resources by different users.

- TDMA

  In Figure 5.2, the time axis is divided into frames and further into slots. Each slot is assigned exclusively to one terminal/user and the terminals occupy the same frequency band in the same cell. A channel in a TDMA system is then referred to as the slot position in frames.

  Example 14 *GSM uses a TDMA type of access method. One frame is approximately 4.16 (126/20) ms and contains 8 time slot; Each time slot* *lasts 0.577 ms and is called a burst in a TDMA frame. A channel is then defined as the slot/burst position in frames. All GSM signals are transmitted using a 200 KHz frequency band. However, a user may jump from one carrier to another on a large time scale using a frequency hopping fashion.*

- CDMA

  In Figure 5.3, different terminals are transmitting using different spreading codes. More than one terminal may share the same frequency and time domains all the time. A channel in a CDMA system is then defined by the spreading code. Ideally, the codes among terminals should be orthogonal so that the receiver can detect the signal addressed to it in the presence of the interference from other users. However, in practice, this is very difficult to realize. The non-orthogonality of codes gives rise to the multiple access interference (MAI) at the receiver side. Sophisticated receiver structures, e.g., multiuser detection (MUD), are needed to eliminate such interference [4].

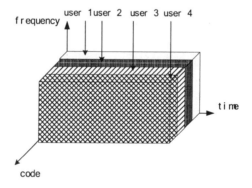

Figure 5.3: CDMA channelization

Example 15 *The IS-95 standard employs CDMA type of access method. In downlink, channels are identified by the orthogonal Walsh code of 64-chip length. Among the 64 channels, one is used as a pilot channel, seven are used as paging channels and 55 are used as traffic channels. The signal is further spread by a 32768-chip m-sequence to distinguish the BSs that are using the same frequency [5].*

- FDMA

  In Figure 5.4, the whole frequency spectrum is divided into several frequency bands, and each terminal is transmitting on a separate band. A channel in an FDMA system is then defined by the frequency band.

Example 16 *The first generation of wireless communication system, Analog American Mobile Phone System (AMPS), uses FDMA as the access method. The allocated frequency band is divided into 30 KHz channels and a call utilizes one 30 KHz channel.*

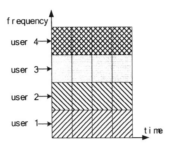

Figure 5.4: FDMA channelization

- OFDMA

  Before the digital implementation of OFDM, FDMA can only be realized using multiple analog RF modules if one terminal is occupying multiple frequency bands. FDMA therefore was deemed unsuitable for broadband communications. However, the rise of OFDM, and in particular, its IFFT/FFT implementation, give FDMA a new life as a broadband multiple access scheme. The use of IFFT/FFT allows terminals to arbitrarily combine multiple frequencies (subcarriers) at the baseband, leading to orthogonal frequency division multiple access (OFDMA) scheme as shown in Figure 5.5.

Definition 1 *An OFDMA system is defined as one in which each terminal occupies a subset of subcarriers (termed an OFDMA traffic channel), and each traffic channel is assigned exclusively to one user at any time [6].*

In OFDMA, users are not overlapped in frequency domain at any given time. However, the frequency bands assigned to a particular user may change over the time as shown in Figure 5.5.

Example 17 *The IEEE 802.16a-e has an OFDMA mode with bandwidth options of either 1.25, 5, 10 or 20 MHz. Depending on the bandwidth, the entire spectrum is divided into 128, 512, 1024 or 2048 subcarriers. For example, a 20 MHz band with a 2048-FFT yields a subcarrier spacing of 9.8 KHz [7]. In time domain, the resource is further devided into frames and subframes that can be allocated to different users.*

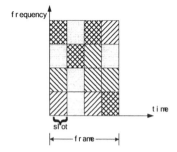

Figure 5.5: OFDMA in frequency and time domain

Up till now, we have introduced two categories of multiple access protocols in wireless systems: contention and non-contention based protocols. It should be pointed out that, in practice, different multiple access protocols may coexist in the same system. For example, a terminal in IS-95 network utilizes ALOHA to initiate the random access process on a random access channel, which is shared by all the terminals. Once its requesting signal is received correctly by the BS, the BS shall assign a dedicated channel to the terminal. Since IS-95 is a CDMA system, the channel will indicate the code that the terminal should use. The data is then exchanged using the allocated channel.

Example 18 *Figure 5.6 illustrates the combination of the two types of protocols in IS-95 networks. Before the terminal establishes a call, it first goes through the random access procedure to request a dedicated channel using ALOHA protocol. Once its request is correctly received by the BS, the BS assigns a dedicated channel to the terminal. The assigned channels are sent to the terminal in a paging channel.*

5.3   OFDMA advantages

Applying the three types of channelization schemes to OFDM network, we arrive at three types of systems: OFDM-TDMA, OFDM-CDMA (MC-CDMA or Multicarrier DS-CDMA) and OFDMA. The flexibility of OFDM-TDMA and OFDM-CDMA lies in the fact that the number of slots or the number of codes assigned to each user is adjustable, leading to different data rates. On the other hand, OFDMA is fundamentally advantageous over OFDM-TDMA and OFDM-CDMA when it comes down to real system operations.

- Granularity: Early broadband access systems utilize OFDM-TDMA to offer a straightforward way of multiple-accessing: each user uses a small number of OFDM symbols in a time slot and, multiple users share the

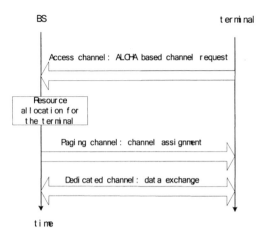

Figure 5.6: The coexistence of contention based and non-contention based MAC schemes in IS-95 system

radio channel through TDMA. The method has two obvious shortcomings. First, every time a user utilizes the channel, it has to burst its data over the entire bandwidth, leading to a high peak power and therefore low RF efficiency. Second, when the number of sharing users is large, the TDMA access delay can be excessive. OFDMA is a much more flexible and powerful way to achieve multiple-access with OFDM modem. In OFDMA, the multiple-access is not only supported in the time domain, but also in the frequency domain, just like traditional FDMA minus the guard-band overhead. As a result, an OFDMA system can support more users with much less delay. The finer data rate granularity in OFDMA, as illustrated in Figure 5.7, is paramount to rich media applications with diverse QoS requirements.

- Link budget: Since each TDMA user must burst its data over the entire bandwidth during the allocated time slots, the instantaneous transmission power (dictated by the peak rate) is the same for all users, regardless of their actual data rates. This inevitably creates a link budget deficit that handicaps the low rate users. Unlike TDMA, an OFDMA system can accommodate a low-rate user by allocating only a small portion of its band (proportional to the requested rate). For example, by reducing the effective transmit bandwidth to 1/64 of the system bandwidth, OFDMA can provide an about 18 dB uplink budget advantage over OFDM-TDMA.

- Receiver simplicity: OFDMA has the merit of easy decoding at the receiver side, as it eliminates the intra-cell interference avoiding CDMA type of multi-user detection. This is not the case in MC-CDMA, even if

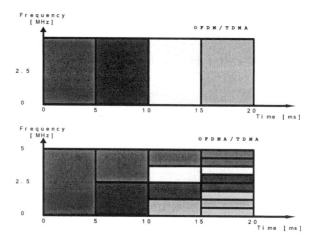

Figure 5.7: Resource partitioning in OFDMA

the codes are designed to be orthogonal. Users' signals can only be detected jointly since the code orthogonality is destroyed by the frequency selective fading. The fact that users' channel characteristics must be estimated also favors OFDMA. In MC-CDMA, users' channel responses must be estimated using complex jointly estimation algorithms. Furthermore, OFDMA is the least sensitive multiple access scheme to system imperfections [9]. Due to these intrinsic features, OFDMA has been adopted in several modern wireless systems, e.g., IEEE 802.16a-e [7], and IEEE 802.20.

- Multiuser diversity: As stated in Chapter 2, broadband signals experience frequency selective fadings. The frequency response of the channel varies over the whole frequency spectrum. The fact that each user has to transmit its signal over the entire spectrum in OFDM-TDMA/CDMA leads to an averaged-down effect in the presence of deep fading and narrowband interference. On the other hand, OFDMA allows different users to transmit over different portions of the broadband spectrum (traffic channel). Since different users perceive different channel qualities, a deep faded channel for one user may still be favorable to others. Therefore, through judicious channel allocation, the system can potentially outperform interference-averaging techniques by a factor of 2 to 3 in spectrum efficiency [8].

Clearly, resource allocation plays an important role in OFDMA systems. In the ensuing sections, we discuss MAC layer issues in OFDMA systems related to resource allocation. We first study a key factor, multiuser diversity, in a multiuser wireless communication environment. Next we establish the OFDMA optimality using optimal resource allocation. Finally, we discuss a cross layer design concept in modern OFDMA systems.

5.4 Multiuser diversity

Multiuser diversity is a recently identified diversity in multiuser wireless communication systems. In general, multiuser diversity gain arises from the fact that in a wireless system with many users, the utility value (e.g., achievable data rate) of a given resource unit varies from one user to another. Such fluctuations allow the overall system performance to be maximized by allocating each radio resource unit to the user that can best exploit it. To illustrate the multiuser diversity gain, let us study a simple example as follows.

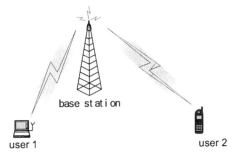

Figure 5.8: A two-user wireless communication system

Example 19 *Figure 5.8 shows a single cell OFDMA system with one base station serving two users. In this example, we have the following assumptions:*

1. *The two users are independent, i.e., their channel responses are independent.*

2. *The users have perfect knowledge of the channel state information.*

3. *There is a perfect feedback channel from each user to the BS which is error free and of no delay; In cases when the uplink and downlink channels are symmetric, e.g., time-division-duplex (TDD) systems, the feedback channel can be omitted since the measurements of the uplink channels by the BS can be used as the estimates of the downlink channels.*

4. *The BS gathers the channel measurements from the two users and allocates channels (i.e., subcarriers) based on these measurement reports.*

Since the SINRs across subcarriers characterise the channel response, we shall use the system average SINR to illustrate the multiuser diversity. Figure

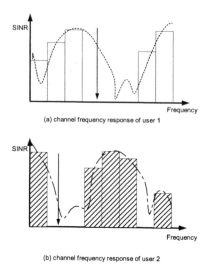

(a) channel frequency response of user 1

(b) channel frequency response of user 2

Figure 5.9: Channel frequency responses of the two users

(a) and (b) in Figure 5.9 show the frequency selective channel responses of two users respectively. Due to interference and noise, some of the subcarriers are in deep fading. However since the two users are independent, a deep-faded subcarrier for one user may be an excellent one for the other user. If we use TDMA, then the SINR value on each subcarrier is the average of the two users, which is indicated in Figure 5.10 (a). On the other hand, if we use OFDMA with intelligent resource allocation, then each subcarrier can be utilized more efficiently by allocating it to the user that has the highest channel frequency response, which is indicated in Figure 5.10 (b). In Figure 5.10 (a) and Figure 5.10 (b), the bold curves in the figures represent the achieved SINR values on subcarriers.

To analyze the multiuser diversity gain numerically, we extend the above example to a K-user case and calculate the achievable transmission rate on a randomly selected subcarrier n with and without resource allocation. Assume unit noise plus interference variance, then $SINR_i$ denotes the channel frequency response of user i on this subcarrier. Assume that all users have the same distribution of $SINR_i$ and denote the CDF and pdf of $SINR_i$ as $F(x)$ and $f(x)$, respectively. We derive the achievable rate with and without intelligent resource allocation as follows.

- Achievable transmission rate with intelligent resource allocation

With intelligent resource allocation, each subcarrier is assigned to the user that has the highest $SINR$ value (denote it as $SINR_{max}$) and the user is denoted as

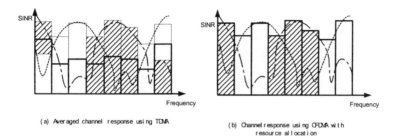

(a) Averaged channel response using TDMA

(b) Channel response using OFDMA with resource allocation

Figure 5.10: Achieved SNR levels in (a) TDMA and (b) OFDMA with resource allocation

$$k_n = \arg\max_{i=1:K}(SINR_i). \tag{5.1}$$

The CDF of $SINR_{\max}$ is then expressed as

$$
\begin{aligned}
F_{SINR_{\max}}(x) &= \Pr(SINR_{\max} < x) \\
&= \Pr(SINR_1 < x)\Pr(SINR_2 < x)..\Pr(SINR_K < x) \\
&= (F(x))^K \\
&\leq F(x) .
\end{aligned}
$$

The last inequality comes from the fact that $0 \leq F(x) \leq 1$. The equality holds if and only if $K = 1$. The pdf of $SINR_{\max}$ is then calculated as

$$f_{SINR_{\max}}(x) = K\,(F(x))^{K-1}\,f(x).$$

Assume that the system employs an adaptive coding and modulation scheme whose rate-SINR function can be expressed as

$$r = g(SINR),$$

where $g(\cdot)$ is a continuous non-decreasing function, i.e., the increase of the SINR always leads to equal or higher transmission rate. An upper bound of $g(\cdot)$ is given by the channel capacity expression with complex channel gain [8]:

$$g(\cdot) = \log_2(1 + \cdot).$$

Then the maximum achievable transmission rate out of the K users, denoted as $r_{\max}^{(K)}$, on this subcarrier is

$$r_{\max}^{(K)} = g(SINR_{\max}).$$

The expected achievable rate is then

$$
\begin{aligned}
E\{r_{\mathrm{max}}^{(K)}\} &= E\{g(SINR_{\mathrm{max}})\} \\
&= \int_0^\infty g(x) f_{SINR_{\mathrm{max}}}(x) dx \\
&= g(x) F_{SINR_{\mathrm{max}}}(x)\big|_0^\infty - \int_0^\infty \frac{dg(x)}{dx} F_{SINR_{\mathrm{max}}}(x) dx \\
&= G - \int_0^\infty \frac{dg(x)}{dx} F_{SINR_{\mathrm{max}}}(x) dx \\
&= G - \int_0^\infty \frac{dg(x)}{dx} \left(F(x)\right)^K dx,
\end{aligned}
\tag{5.2}
$$

where G is the maximum achievable rate that the system allows, e.g., the maximum ACM order supported.

- Achievable transmission rate without resource allocation

On the other hand, without resource allocation, the subcarrier is assigned to a randomly selected user, then the achievable rate on this subcarrier is

$$
r_{\mathrm{rand}} = g(SINR_i).
$$

The expected rate is calculated as

$$
\begin{aligned}
E\{r_{\mathrm{rand}}\} &= E\{g(SINR_i)\} \\
&= \int_0^\infty g(x) f(x) dx \\
&= g(x) F(x)\big|_0^\infty - \int_0^\infty \frac{dg(x)}{dx} F(x) dx \\
&= G - \int_0^\infty \frac{dg(x)}{dx} F(x) dx.
\end{aligned}
\tag{5.3}
$$

The gap between (5.2) and (5.3) is then expressed as

$$
\begin{aligned}
\triangle r &= E\{r_{\mathrm{max}}^{(K)}\} - E\{r_{\mathrm{rand}}\} \\
&= \int_0^\infty \frac{dg(x)}{dx} \left(F(x) - \left(F(x)\right)^K\right) dx.
\end{aligned}
$$

Since $g(\cdot)$ is non-decreasing, $\frac{dg(x)}{dx} \geq 0$. Also notice that $F(x) - \left(F(x)\right)^K \geq 0$, therefore we conclude that

$$
E\{r_{\mathrm{max}}^{(K)}\} \geq E\{r_{\mathrm{rand}}\}.
\tag{5.4}
$$

The equality holds if and only if $K = 1$. It is also obvious that

$$
E\{r_{\mathrm{max}}^{(K)}\} > E\{r_{\mathrm{max}}^{(K-1)}\}... > E\{r_{\mathrm{max}}^{(1)}\} = E\{r_{\mathrm{rand}}\}
\tag{5.5}
$$

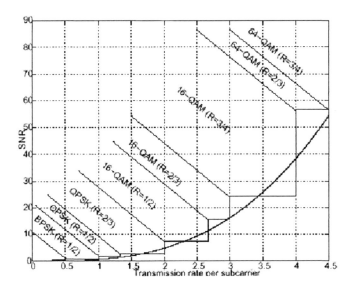

Figure 5.11: Adaptive coding and modulation for BER $P_e = 1 \times 10^{-6}$

Thus Δr is a monotonically increasing function with respect to K.
(5.4) and (5.5) imply that intelligent resource allocation renders performance improvement, and the improvement increases with the number of users.

Definition 2 *The system performance improvement (in the above case, the performance is evaluated as the expected transmission rate per subcarrier) due to the increase in the number of users is referred to as "multiuser diversity gain".*

The multiuser diversity gain is attributed to users' randomness and independence. With intelligent resource allocation, the resource unit can be assigned to the certain user that bears a high utility value. Increasing the number of users results in a higher probability of finding such a user, leading to the performance improvement. To appreciate the multiuser diversity gain numerically, let us study a concrete example.

Example 20 *Let $SINR_i$ be i.i.d complex Gaussian random variable with zero mean and unit variance. Let the ACM be characterized by the rate-SNR/SINR function shown in Figure 5.11 and the rate-SINR/SNR function be approximated as $r = g(SINR_i) = \left(\frac{SNR_i}{0.6}\right)^{1/3}$. The SINR values with respect to the number of users are plotted for cases with resource allocation ($SINR_{max}$) and random assignment ($SINR_i$) in Figure 5.12. The corresponding achievable transmission rates, using (5.2) and (5.3), are plotted in Figure 5.13. The results indicate that as the number of users increases, the gap between the two rates (with and without*

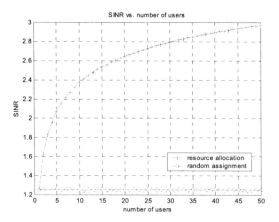

Figure 5.12: SINR vs. K for intelligent resource allocation and random assignment

resource allocation) widens. The gap then quantifies the multiuser diversity gain obtained from the intelligent resource allocation, and the gain increases with the number of users.

Multiuser diversity arises naturally from the multiuser communication environment. In the above example, the diversity enriched in users' independent channel characteristics is exploited by the resource allocations. Similarly, the diversity embedded in users' independent traffic variations can be utilized as well. For example, with bursty traffic, some users may have no packets to send for a period of time while others may be heavily loaded. A good resource allocation scheme should assign the channel to the user that has traffic to send as well as has good channel condition. As the number of users increases, the probability of finding such a user increases too [11].

The multiuser diversity gain is also available in wireless relay networks. Grossglauser and Tse investigated the mobility-enabled multiuser diversity in relay networks [12]. The basic idea is that if the number of nodes in the relay network is very large and the nodes have independent mobility, the probability that at least one node can relay the information from the source node to the destination node successfully is significant.

Another example of multiuser diversity gain is given in [13] where the space-enabled multiuser diversity is exploited in a space-division multiple access (SDMA) system. A five-user case is illustrated in Figure 5.14.

Example 21 *The left figure in Figure 5.14 shows the spatial signatures of five users. User 1 and user 3 have near-orthogonal spatial signatures. User 2 and user 4 have near-orthogonal spatial signatures. The orthogonality between the spatial signatures implies that the mutual interference from the other terminal*

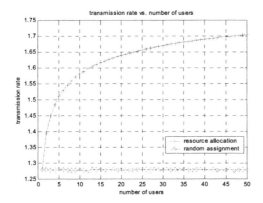

Figure 5.13: Transmission rate vs. K for intelligent resource allocation and random assignment

in each pair is negligible if assigned to the same channel. Using random assignment, as shown in the upper right assignment in Figure 5.14, user 1 and user2 are assigned to the same channel. Since the two users' signatures have high correlation, they cause strong interference to each other, leading to degraded performance. On the other hand, with judicious channel-aware assignment, as indicated in the bottom right assignment in Figure 5.14, the users with orthogonal spatial signatures are assigned to the same channel, eliminating interference from each other. It is obvious that the channel-aware assignment yields better performance as less interference is introduced.

It has been proved that an SDMA system approaches its capacity using intelligent resource allocation strategies with a large number of users [13]. This is achieved by packing as many users as possible on the same channel as long as they have orthogonal spatial signatures. It is obvious that, if users are independent, their spatial signatures shall be randomly located in the spatial signature plane. As a consequence, with the increase of users, the number of user tuples that have orthogonal spatial signatures increases too. By intelligent resource allocation, the system throughput can be improved by assigning these user tuples to the same channel.

Multiuser diversity has attracted increasing attention in MAC layer design of modern wireless networks especially in the OFDMA system due to the enriched multiuser frequency selectivity. In the rest of this chapter, we discuss some MAC layer related design issues for OFDMA systems including multiuser diversity exploitation. In particular, we establish the optimality of OFDMA in multiuser multicarrier SISO and MIMO systems. This is accomplished by solving the optimal power/subcarrier allocation problems in the generic multiuser multicarrier SISO and MIMO system models. In addition, we propose two simplified resource allocation schemes for OFDMA/MIMO systems which

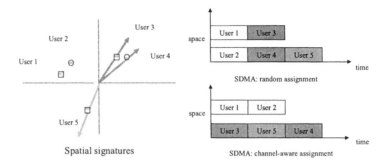

Figure 5.14: Spatial multiuser diversity.

approach the optimal solution in low and high SNR regions, respectively.

5.5 OFDMA optimality

Recently, many broadband wireless networks have included the MIMO option in their protocols. Compared to the SISO system, MIMO offers better resistance against fading. In addition, the higher diversity can potentially lead to a multiplicative increase in capacity. Please refer to [14]-[23] and therein for more references on MIMO. In principle, OFDMA and MIMO can be synergistically integrated to offer the benefits of both system simplicity and high performance. In fact, such has been adopted in the IEEE 802.16 standard. Despite these promises, a few fundamental questions remain as whether or not OFDMA/MIMO is the right choice for multiuser MIMO systems. Even the OFDMA optimality in multiuser multicarrier SISO systems has not been fully established. [24] proves the optimality of OFDMA in a downlink multiuser multicarrier SISO system with adaptive QAM modulation and independent decoding. However, the proof cannot be generalized to other modulation schemes.

In this section, we answer the following questions:

1. Is OFDMA the optimal choice for multiuser multicarrier SISO systems?

2. Is OFDMA the optimal choice for multiuser multicarrier MIMO systems?

We first provide a proof of the OFDMA/SISO optimality from the information theory point of view for the downlink multiuser multicarrier system with independent decoding. Furthermore, we show that the optimality of OFDMA holds for any adaptive modulation scheme whose transmission rate can be approximated as a convex function in terms of SNR/SINR. Next, we extend the OFDMA optimality to a generic multiuser multicarrier MIMO system model. We show that OFDMA/MIMO is the optimal downlink scheme under the independent decoding constraint. We also derive a set of joint conditions under

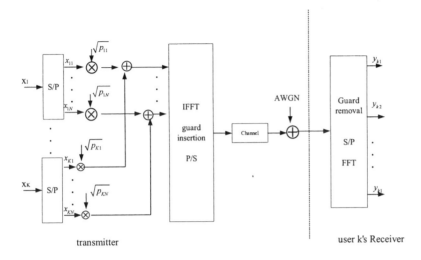

Figure 5.15: The multiuser multicarrier SISO downlink system model

which the optimal subcarrier and the optimal power allocation should be performed. However, unlike OFDMA/SISO where the optimal solution can be found with linear complexity (with respect to the number of users and the number of the subcarriers), the OFDMA/MIMO problem has no explicit solution due to the tangled effects of subcarrier allocation and power loading. To make the problem tractable, we propose two suboptimal subcarrier allocation criteria. The two criteria approach the optimal solution in low SNR and high SNR regions, respectively. Besides, the two criteria result in linear complexity with respect to the number of users and the number of subcarriers in resource allocations.

5.5.1 Multiuser multicarrier SISO systems

System model

We consider the downlink multiuser multicarrier SISO system model as shown in Figure 5.15. We invoke the following assumptions:

- The transmitted signals experience slow Rayleigh fading channels between the transmitter and the receiver. As a result, the channel coefficients can be regarded as constants in resource allocation;

- The transmitter has perfect knowledge of the channel state information.

Let the number of users be K, the number of subcarriers be N and the channel gain associated with subcarrier n and user k be h_{kn}. For a generic multicarrier system, we assume that all the users can transmit on all the subcarriers and the total power is constrained to Q :

$$\sum_{n=1}^{N}\sum_{k=1}^{K} p_{kn} = Q,$$

where $\sqrt{p_{kn}}$ is the transmission power scalar of user k on subcarrier n. We will show shortly that the optimal power allocation scheme allows only one user to have nonzero transmission power on each subcarrier.

The transmitted signal on subcarrier n is $\sum_{i=1}^{K} x_{in}\sqrt{p_{in}}$, and the received signal by user k on subcarrier n is expressed as:

$$
\begin{aligned}
y_{kn} &= \left(\sum_{i=1}^{K} x_{in}\sqrt{p_{in}}\right) h_{kn} + v_{kn} \\
&= x_{kn}\sqrt{p_{kn}}h_{kn} + \left(\sum_{i\neq k}^{K} x_{in}\sqrt{p_{in}}\right) h_{kn} + v_{kn},
\end{aligned}
\tag{5.6}
$$

where x_{in} is the transmitted signal from the base station to user i on subcarrier n and v_{kn} is the AWGN noise with variance N_0. With independent decoding, the signal to noise and interference ratio (SINR) perceived by user k on subcarrier n is expressed as (assuming $E\left(\|x_{in}\|^2\right) = 1$ and x_{in}, x_{jn} are identical independent complex Gaussian random variables with zero means):

$$
SINR_{kn} = \frac{p_{kn}\|h_{kn}\|^2}{\displaystyle\sum_{\substack{i=1\\i\neq k}}^{K} p_{in}\|h_{kn}\|^2 + N_0}.
\tag{5.7}
$$

The capacity achieved by user k on subcarrier n without the knowledge of other users' information is given by [8]

$$
c_{kn} = \log_2\left(1 + \frac{p_{kn}\|h_{kn}\|^2}{\displaystyle\sum_{\substack{i=1\\i\neq k}}^{K} p_{in}\|h_{kn}\|^2 + N_0}\right).
$$

The total capacity over all subcarriers and all users is then

$$
C\left([p_{kn}]_{K\times N}\right) = \sum_{n=1}^{N}\sum_{k=1}^{K}\log_2\left(1 + \frac{p_{kn}\|h_{kn}\|^2}{\displaystyle\sum_{\substack{i=1\\i\neq k}}^{K} p_{in}\|h_{kn}\|^2 + N_0}\right).
$$

In order to maximize the total capacity, the transmission power needs to be distributed optimally under the total power constraint. We formulate the following optimization problem to determine the optimal power allocation. It turns out that under the optimal power allocation, only one user is transmitting on each subcarrier, i.e., OFDMA.

$$\max \; C\left([p_{kn}]_{K \times N}\right) \;=\; \sum_{n=1}^{N}\sum_{k=1}^{K} \log_2 \left(1 + \frac{p_{kn}\left\| h_{kn} \right\|^2}{\sum_{\substack{i=1 \\ i \neq k}}^{K} p_{in}\left\| h_{kn} \right\|^2 + N_0} \right)$$

$$s.t. \quad \sum_{n=1}^{N}\sum_{k=1}^{K} p_{kn} = Q \tag{5.8}$$

$$p_{kn} \geq 0, k = 1, ...K; n = 1, ..., N.$$

Note that if the optimal solution to (5.8) suggests $p_{kn} = 0$, then subcarrier n should not be assigned to user k. Therefore there is also an embedded subcarrier allocation in this problem.

The optimality of SISO/OFDMA

The following theorem states that the optimal solution to (5.8) is OFDMA.

Theorem 2 C *is maximized if the following conditions are satisfied:*

1. *Each subcarrier is assigned to only one user, i.e., OFDMA;*

2. *The assigned user on subcarrier n has the highest channel gain over K users.*

3. *The power over subcarriers is allocated using a water-filling solution with respect to channel gains* $(h_{\max,1}, h_{\max,2}, , ..., h_{\max,N})$, *where $h_{\max,n}$ is the maximum channel gain on subcarrier n over K users.*

The proof of Theorem 2 uses the following definitions and Lemma:

Definition 3 *The domain of a function $f(\mathbf{x})$: $R^n \to R$, dom f, is defined as* $\{\mathbf{x} | f(\mathbf{x}) < \infty\}$.

Definition 4 *A function f: $R^n \to R$ is convex if the domain of f (dom f) is a convex set and if for $\mathbf{x}, \mathbf{y} \in$ dom f, and $\theta \in [0, 1]$, we have*

$$f(\theta\mathbf{x} + (1 - \theta)\mathbf{y}) \leq \theta f(\mathbf{x}) + (1 - \theta)f(\mathbf{y}).$$

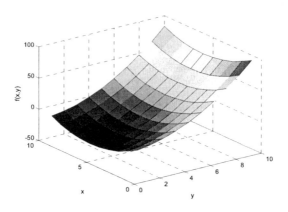

Figure 5.16: An example of a convex function with two variables.

Lemma 1 *[25, Corollary 32.3.4] Let f be a convex function, and let S be a non-empty polyhedral convex set contained in dom f. If S contains no lines (infinite), and f is bounded above on S, then the supremum (maximum) of f relative to S is attained at one of the finitely many extreme points of S (i.e., vertices of S).*

Example 22 *A two dimensional example explaining Lemma 1 is shown in Figure 5.16. Here $f(x,y) = x^2 + y^2 - 12x + 3$ and $S = \{(x,y)|x, y \in [0,10]\}$. S is a polyhedral convex set (compact and bounded) and $f(x,y)$ is easily proved to be convex on S. Then the maximum of $f(x,y)$ with respect to S is attained at one of S's vertices. The vertices of S are: (0,0), (0,10), (10,0) and (10,10). It is then obvious to see that the maximum of $f(x,y)$ is achieved at (0,10).*

We now prove the first two claims in Theorem 2 using Lemma 1.

Proof. Assume that q_n amount of power has been assigned to subcarrier n (of course, $\sum\limits_{n=1}^{N} q_n = Q$), then (5.8) can be re-formulated as follows

$$\max \; C\left([p_{kn}]_{K\times N}\right) \;=\; \sum_{n=1}^{N} C_n\left(p_{1n},...,p_{Kn}\right) \tag{5.9}$$

$$s.t. \qquad C_n\left(p_{1n},...,p_{Kn}\right) = \sum_{k=1}^{K} \log_2 \left(1 + \frac{p_{kn}\,\|h_{kn}\|^2}{\sum\limits_{\substack{i=1\\i\neq k}}^{K} p_{in}\,\|h_{kn}\|^2 + N_0} \right)$$

$$\sum_{k=1}^{K} p_{kn} = q_n, n = 1,...,N$$

$$p_{kn} \geq 0, k = 1,...K; n = 1,...,N$$

$$\sum_{n=1}^{N} q_n = Q$$

$$q_n \geq 0, n = 1,...,N. \tag{5.10}$$

(5.9) can be further decoupled into parallel optimization problems for different subcarriers. On subcarrier n, the optimization problem is expressed as:

$$\max \qquad C_n(p_{1n},...,p_{Kn}) = \sum_{k=1}^{K} \log_2 \left(1 + \frac{p_{kn}\,\|h_{kn}\|^2}{\sum\limits_{\substack{i=1\\i\neq k}}^{K} p_{in}\,\|h_{kn}\|^2 + N_0} \right) \tag{5.11}$$

$$s.t. \qquad \sum_{k=1}^{K} p_{kn} = q_n$$

$$p_{kn} \geq 0, \; k = 1,...K.$$

It is clear that if C_n is maximized for all n, then $C = \sum\limits_{n=1}^{N} C_n$ is maximized in (5.9). So let us focus on the optimization problem of a specific subcarrier n. We now drop some of the subscript n for simplicity, then the sub-problem can be re-written as:

$$\max \qquad C_n(\underbrace{p_1,...,p_K}_{\mathbf{p}}) = \sum_{k=1}^{K} \log_2 \left(1 + \frac{p_k}{q_n - p_k + N_0/\|h_k\|^2} \right) \tag{5.12}$$

$$s.t. \qquad \sum_{k=1}^{K} p_k = q_n \tag{5.13}$$

$$p_k \geq 0, \; k = 1,...K. \tag{5.14}$$

The feasible region of (5.12) is

$$S = \left\{ (p_1, ..., p_K) \in R^K \mid \sum_{k=1}^{K} = q_n, p_k \geq 0, k = 1, ...K \right\}.$$

This is a bounded polyhedral (i.e., a polytope) defined by linear equality (5.13) and linear inequalities (5.14). It has K vertices with the kth vertex being $\{\mathbf{p} \in R^K \mid p_k = q_n$ and $p_i = 0$ for $i \neq k\}$. It can be easily proved that $C_n(\mathbf{p})$ is a convex function. (Please see Appendix I for the proof.) Then according to Lemma 1, the maximum point of C_n is achieved at one of S's vertices. In this case, the maximum point is achieved when only one element of $\mathbf{p}$ is nonzero. It is then trivial to verify that the only nonzero position of $\mathbf{p}$ corresponds to the user that has the highest channel gain, i.e., $argmax\{p_1, ...p_K\} = argmax\{h_1, ...h_K\}$. As a result, we have proved that on a specific subcarrier, the optimal power allocation allows only one user to transmit with all the power assigned to this subcarrier and the user has the highest channel gain on that subcarrier. Note that with the optimal power allocation on subcarrier n,

$$C_n(p_1, ..., p_K) = \log_2\left(1 + \frac{q_n}{N_0} \|h_{\text{max},n}\|^2\right), \tag{5.15}$$

where

$$\|h_{\text{max},n}\|^2 = \max(\|h_{1n}\|^2, ..., \|h_{Kn}\|^2).$$

This finishes the proof of the first two claims in Theorem 2.   ∎

Remark 3 *Note that the proof of the OFDMA optimality in Theorem 2 can be generalized to any convex rate-SNR/SINR function besides the information capacity expression. For example, the function used in [24] is $C = \log_2(1 + \frac{SINR}{\Gamma})$, where Γ is an implementation factor corresponding to QAM. It can be easily verified that this is a convex function. Therefore with Lemma 1, the OFDMA optimality can be easily established for the system model used in [24].*

The proof of the third claim in Theorem 2 uses the following Lemma.

Lemma 2 *Karush-Kuhn-Tucker (KKT) conditions (First-Order Necessary Conditions) [26, Theorem 12.1]*

For a general optimization problem

$$\min_{\mathbf{x} \in R^n} f(\mathbf{x}) \tag{5.16}$$

$$s.t. \quad g_i(\mathbf{x}) = 0, \ i \in \mathcal{E}$$

$$g_i(\mathbf{x}) \geq 0, \ i \in \mathcal{I}$$

where $f(\mathbf{x})$ and $g_i(\mathbf{x})$ are all smooth, real-valued functions on a subset of R^n. $\mathcal{I}$ and $\mathcal{E}$ are two finite sets of indices. The Lagrangian of (5.16) is defined as

$$L(\mathbf{x}, \boldsymbol{\mu}) = f(\mathbf{x}) - \sum_{i \in \mathcal{E} \cup \mathcal{I}} \mu_i g_i(\mathbf{x}).$$

If $\mathbf{x}^*$ is a local solution of (5.16) and the linear independence constraint qualification (LICQ) holds at $\mathbf{x}^*$, then there is a Lagrangian multiplier vector $\boldsymbol{\mu}^*$ with components $\mu_i^*, i \in \mathcal{E} \cup \mathcal{I}$, such that the following conditions are satisfied at $(\mathbf{x}^*, \boldsymbol{\mu}^*)$:

$$\nabla_{\mathbf{x}} L(\mathbf{x}^*, \boldsymbol{\mu}^*) = 0 \tag{5.17}$$

$$g_i(\mathbf{x}^*) = 0, \ i \in \mathcal{E} \tag{5.18}$$

$$g_i(\mathbf{x}^*) \geq 0, \ i \in \mathcal{I} \tag{5.19}$$

$$\boldsymbol{\mu}^* \geq 0, \ i \in \mathcal{I} \tag{5.20}$$

$$\mu_i^* g_i(\mathbf{x}^*) = 0, \ i \in \mathcal{E} \cup \mathcal{I} \tag{5.21}$$

Furthermore, if (5.16) is a convex optimization problem (the objective function $f(\mathbf{x})$ is convex and the feasible region is also convex), then conditions from (5.17) to (5.21) are also sufficient conditions to determine the global optimal solution $\mathbf{x}^*$ [27].

We now prove the third claim in Theorem 2.

Proof of the third claim of Theorem 2.   Referring to (5.15), (5.11) now can be re-written as:

$$\min \quad C(q_1, ...q_N) = \sum_{n=1}^{N} \log_2 \left(1 + \frac{q_n}{N_0} \|h_{\max,n}\|^2 \right) \tag{5.22}$$

$$s.t. \quad \sum_{n=1}^{N} q_n = Q$$

$$q_n \geq 0, n = 1, 2, ..., N.$$

It can be easily proved to be a convex optimization problem and that the LICQ condition holds. Therefore we can use the KKT conditions in Lemma 2 to determine the optimal solution. In order to do so, we first construct the Lagrangian of (5.22) as

$$L(q_1, ...q_N, \alpha, \beta_1, ..., \beta_N) = -\sum_{n=1}^{N} \log_2 \left(1 + \frac{q_n}{N_0} \|h_{\max,n}\|^2 \right)$$
$$-\alpha \left(\sum_{n=1}^{N} q_n - Q \right) - \sum_{n=1}^{N} \beta_n q_n.$$

Using the KKT condition, the optimal solution $\mathbf{q}^* = (q_1^*, ..., q_N^*)$ satisfies the following conditions:

$$\nabla_{\mathbf{q}} L(\mathbf{q}^*, \alpha^*, \beta_1^*, ..., \beta_N^*) = 0 \tag{5.23}$$

$$\sum_{n=1}^{N} q_n^* - Q = 0.$$

$$q_n^* \geq 0$$

$$\beta_n^* \geq 0, \ n = 1, ..., N$$

$$\beta_n^* q_n^* = 0, \ n = 1, ..., N. \tag{5.24}$$

With (5.23) we have

$$\frac{\partial L}{\partial q_n}(\mathbf{q}^*, \alpha^*, \beta_1^*, ..., \beta_N^*) = -\frac{1}{q_n^* + \frac{N_0}{\|h_{\max,n}\|^2}} - \alpha^* - \beta_n^* = 0. \qquad (5.25)$$

Using (5.24), if $q_n^* > 0$, then $\beta_n^* = 0$. Following (5.25), if $q_n^* > 0$ then it satisfies the following condition:

$$\frac{1}{q_n^* + \frac{N_0}{\|h_{\max,n}\|^2}} + \alpha^* = 0, \quad n = 1, 2, ...N. \qquad (5.26)$$

As a result, the optimal power distribution across subcarriers is

$$q_n^* = \left(-\frac{1}{\alpha} - \frac{N_0}{\|h_{\max,n}\|^2}\right)^+, \qquad (5.27)$$

where $(\cdot)^+ = max(0, \cdot)$ and α satisfies $\sum_{n=1}^{N} q_n = Q$. $-\frac{1}{\alpha}$ is then called the "water-level" of the solution to (5.22). This is well known as the "water-filling" solution in [28] for parallel channels. ∎

5.5.2   Multiuser multicarrier MIMO systems

In this section, we extend our study on the OFDMA optimality to MIMO systems. We show that OFDMA/MIMO is the optimal downlink scheme under the independent decoding constraint. We also derive a set of joint conditions under which the optimal subcarrier and the optimal power allocation should be performed. To reduce the loading complexity, we propose two suboptimal subcarrier allocation criteria. The two low-complexity loading criteria approach the optimal solution in low SNR and high SNR region, respectively.

System model

We consider a generic downlink multiuser multicarrier MIMO system model as shown in Figure 5.17. Let the number of transmitter antennas be N_t and the number of receive antennas be N_r. Denote the transmitted signal vector intended for user k on subcarrier n as

$$\mathbf{x}_{kn} = \left(\mathbf{x}_{kn}(1), ..., \mathbf{x}_{kn}(N_t)\right)^T,$$

where $\mathbf{x}_{kn}(t)$ represents the transmitted signal for user k from the tth antenna on subcarrier n. Please see Figure 5.17 for illustrations. The transmitted vector signal on subcarrier n is then expressed as:

$$\sum_{i=1}^{K} \mathbf{x}_{in}\sqrt{p_{in}} = \left(\sum_{i=1}^{K}\sqrt{p_{in}}\mathbf{x}_{in}(1), ..., \sum_{i=1}^{K}\sqrt{p_{in}}\mathbf{x}_{in}(N_t)\right)^T.$$

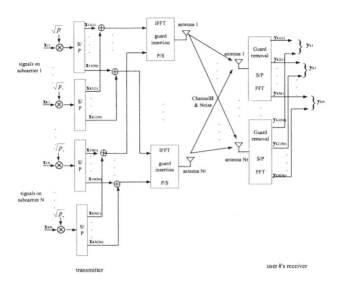

Figure 5.17: The multiuser multicarrier MIMO downlink system model

Now we define the received signal by user k on subcarrier n as

$$\mathbf{y}_{kn} = (\mathbf{y}_{kn}(1), ..., \mathbf{y}_{kn}(N_r))^T .$$

Further denote the channel gain matrix on subcarrier n for user k as $\mathbf{H}_{kn}$ ($N_r \times N_t$) and denote $\mathbf{v}_{kn}$ as the AWGN noise vector. The input-output relationship between $\mathbf{x}_{kn}$ and $\mathbf{y}_{kn}$ is thus

$$\mathbf{y}_{kn} = \mathbf{H}_{kn} \left(\sum_{i=1}^{K} \mathbf{x}_{in} \sqrt{p_{in}} \right) + \mathbf{v}_{kn}.$$

Assume that

$$
\begin{aligned}
E\left\{\mathbf{x}_{kn}\mathbf{x}_{kn}^H\right\} &= \mathbf{I}, \\
E\left\{\mathbf{x}_{in}\mathbf{x}_{jn}^H\right\} &= 0, \text{ if } i \neq j, \\
E\left\{\mathbf{v}_{kn}\mathbf{v}_{kn}^H\right\} &= N_0\mathbf{I}.
\end{aligned}
$$

Again assume q_n is the amount of power assigned to subcarrier n (i.e., $\sum_{i=1}^{K} p_{in} = q_n$). Consequently,

$$\mathbf{R}_{\mathbf{y}_{kn}} = E\left\{\mathbf{y}_{kn}\mathbf{y}_{kn}^H\right\} = \mathbf{H}_{kn}\mathbf{H}_{kn}^H q_n + N_0.$$

With independent decoding, the maximum achievable rate of user k on subcarrier n is given by [29]

$$
\begin{aligned}
c_{kn} &= max\; I\left(\mathbf{x}_{kn}; \mathbf{y}_{kn}\right) \\
&= H(\mathbf{y}_{kn}) - H\left(\mathbf{y}_{kn}|\mathbf{x}_{kn}\right) \\
&= \log\left(2\pi e\right)^K \det\left(\mathbf{R}_{\mathbf{y}_{kn}}\right) \\
&\quad - \log\left(2\pi e\right)^K \det\left(\mathbf{R}_{\mathbf{y}_{kn}} - \mathbf{H}_{kn}\mathbf{H}_{kn}^H p_{kn}\right) \\
&= \log \frac{\det\left(\mathbf{H}_{kn}\mathbf{H}_{kn}^H q_n + N_0\mathbf{I}\right)}{\det\left(\mathbf{H}_{kn}\mathbf{H}_{kn}^H(q_n - p_{kn}) + N_0\mathbf{I}\right)} \\
&= \log \frac{\det\left(\mathbf{I} + \frac{\mathbf{H}_{kn}\mathbf{H}_{kn}^H q_n}{N_0}\right)}{\det\left(\mathbf{I} + \frac{\mathbf{H}_{kn}\mathbf{H}_{kn}^H(q_n - p_{kn})}{N_0}\right)}.
\end{aligned}
$$

In order to maximize the total capacity, the transmission power must be distributed optimally under the total power constraint. Similar to the optimal power allocation problem for SISO system, we formulate the following optimization problem for MIMO system:

$$
max\; C\left([p_{kn}]_{K \times N}\right) = \sum_{n=1}^{N}\sum_{k=1}^{K} \log \frac{\det\left(\mathbf{I} + \frac{\mathbf{H}_{kn}\mathbf{H}_{kn}^H q_n}{N_0}\right)}{\det\left(\mathbf{I} + \frac{\mathbf{H}_{kn}\mathbf{H}_{kn}^H(q_n - p_{kn})}{N_0}\right)} \tag{5.28}
$$

$$
s.t. \quad \sum_{n=1}^{N}\sum_{k=1}^{K} p_{kn} = Q
$$

$$
p_{kn} \geq 0, k = 1, ..., K, n = 1, .., N.
$$

It turns out, same as the conclusion for SISO system, that the optimal power loading scheme allows only one user on each subcarrier.

The optimality of MIMO/OFDMA

In this section, we will prove the following theorem which asserts the optimality of OFDMA/MIMO.

Theorem 4 *Referring to (5.28), the total capacity C is maximized when the following conditions are satisfied:*

1. *Each subcarrier is assigned to only one user, i.e., OFDMA.*

2. *The user assigned to subcarrier n has the highest value of $\prod_{i=1}^{M_{kn}}\left(1 + \frac{\lambda_{kn}^{(i)}q_n^*}{N_0}\right)$*

over all k, i.e.,

$$argmax_k (p_{1n}, .., p_{Kn}) = argmax_k \prod_{i=1}^{M_{kn}} \left(1 + \frac{\lambda_{kn}^{(i)} q_n^*}{N_0} \right) \triangleq k_n, \qquad (5.29)$$

where M_{kn} is the rank of $\mathbf{H}_{kn}$, $\left\{ \lambda_{kn}^{(i)} \right\}_{i=1:M_{kn}}$ are the eigenvalues of $\mathbf{H}_{kn}\mathbf{H}_{kn}^H$ and q_n^* is the optimal power assigned to subcarrier n which satisfies the next condition.

3. *The power distribution over subcarriers is $q_n^* = \max(0, q_n)$ where q_n is the root of the following equations,*

$$\sum_{i=1}^{M_{k_n n}} \frac{\lambda_{k_n n}^{(i)}}{\lambda_{k_n n}^{(i)} q_n + N_0} + \alpha = 0, n = 1, 2, ...N, \qquad (5.30)$$

where k_n is the allocated user index on subcarrier n and α satisfies $\sum_{n=1}^{N} q_n^* = Q$. In case of $N_t = N_r = 1$, the optimal power distribution across subcarriers reduces to the standard water-filling solution:

$$q_n^* = \left(-\frac{1}{\alpha} - \frac{N_0}{\lambda_{k_n}^{(1)}} \right)^+.$$

where $(\cdot)^+ = max(0, \cdot)$ and α satisfies

$$\sum_{n=1}^{N} \left(-\frac{1}{\alpha} - \frac{N_0}{\lambda_{k_n}^{(1)}} \right)^+ = Q.$$

Proof. Similar to the proof of the SISO problem, the optimization problem in (5.28) can be decoupled for different subcarriers, allowing us to focus on a specific subcarrier n. As a consequence, solving (5.28) needs to solve the following sub-problem (we drop some of the subcarrier index n for simplicity). The problem is formulated as how to distribute q_n optimally over all users assuming that the total assigned power on subcarrier n is q_n.

$$\max_{\mathbf{p}} C_n \left(\underbrace{p_1, ..., p_K}_{\mathbf{p}} \right) = \sum_{k=1}^{K} \log \frac{\det \left(\mathbf{I} + \frac{\mathbf{H}_k \mathbf{H}_k^H q_n}{N_0} \right)}{\det \left(\mathbf{I} + \frac{\mathbf{H}_k \mathbf{H}_k^H (q_n - p_k)}{N_0} \right)} \qquad (5.31)$$

$$s.t. \quad \sum_{k=1}^{K} p_k = q_n \qquad (5.32)$$

$$p_k \geq 0, k = 1, ..., K. \qquad (5.33)$$

Let

$$\mathbf{H}_k \mathbf{H}_k^H = \mathbf{U}_k^H \mathbf{\Lambda}_k \mathbf{U},$$

where
$$\mathbf{\Lambda}_k = diag(\lambda_k^{(1)}, ..., \lambda_k^{(M_k)}),$$

$\mathbf{U}$ is a unitary matrix and $M_k = rank(\mathbf{H}_k)$, then C_n can be expressed as:

$$
\begin{aligned}
C_n(p_1, ...p_K) &= \sum_{k=1}^{K} \log \prod_{i=1}^{M_k} \left(1 + \frac{\lambda_k^{(i)} p_k}{\lambda_k^{(i)}(q_n - p_k) + N_0} \right) && (5.34) \\
&= \sum_{k=1}^{K} \underbrace{\sum_{i=1}^{M_k} \log \left(1 + \frac{\lambda_k^{(i)} p_k}{\lambda_k^{(i)}(q_n - p_k) + N_0} \right)}_{\overset{\triangle}{=} c_k(p_k)} && (5.35) \\
&= \sum_{k=1}^{K} c_k(p_k).
\end{aligned}
$$

The feasible region of (5.31), S, is the same as (5.12). Also, it can be proved that $C_n(p_1, ...p_K)$ is a convex function over S. (Please see Appendix II for the proof.) Therefore, following the similar argument, the maximum point of C_n is achieved at one of S's vertices and there should be only one nonzero element in $\mathbf{p}$. In this case, the only nonzero element position of $\mathbf{p}$ corresponds to the user that has the highest $c_k(q_n)$, i.e.,

$$argmax\{p_1, ...p_K\} = argmax\{c_1(q_n), ...c_K(q_n)\} \overset{\triangle}{=} k_n.$$

As a result, we have proved that, for MIMO systems, on each specific subcarrier, the optimal solution allows only one user to transmit with all the power assigned on that subcarrier. Note that with the optimal power load on subcarrier n,

$$C_n(p_1, ..., p_K) = \sum_{k=1}^{M_{k_n}} \log \left(1 + \frac{\lambda_{k_n}^{(i)} q_n}{N_0} \right).$$

We conclude that the optimal power loading or equivalently, the optimal subcarrier allocation criterion for subcarrier n is

$$
\begin{aligned}
k_n &= \arg\max_{k} \; p_k && (5.36) \\
&= \arg\max_{k} \sum_{i=1}^{M_k} \log \left(1 + \frac{\lambda_k^{(i)} q_n}{N_0} \right).
\end{aligned}
$$

This finishes the first two claims of Theorem 4.

To prove the third claim, note that (5.28) now reduces to determining the optimal power loading over all the subcarriers, or equivalently:

$$\min_{q_1,\dots,q_N} \quad -\sum_{n=1}^{N}\left(\sum_{i=1}^{M_{k_n n}} \log\left(1 + \frac{\lambda_{k_n n}^{(i)} q_n}{N_0}\right)\right) \tag{5.37}$$

$$s.t. \quad \sum_{n=1}^{N} q_n = Q$$

$$q_n \geq 0, n = 1, 2, \dots, N.$$

It can be easily proved to be a convex optimization problem and that the LICQ condition holds, therefore we can apply Lemma 2 to (5.37). We first construct the Lagrangian of (5.37) as

$$L(q_1, \dots q_N, \alpha, \beta_1, \dots, \beta_N) = -\sum_{n=1}^{N}\sum_{i=1}^{M_{k_n n}} \log\left(1 + \frac{\lambda_{k_n n}^{(i)} q_n}{N_0}\right)$$

$$-\alpha\left(\sum_{n=1}^{N} q_n - Q\right) - \sum_{n=1}^{N} \beta_n q_n.$$

Using the KKT condition, the optimal solution $\mathbf{q}^* = (q_1^*, \dots, q_N^*)$ satisfies the following conditions:

$$\nabla_{\mathbf{q}} L(\mathbf{q}^*, \alpha^*, \beta_1^*, \dots, \beta_N^*) = 0 \tag{5.38}$$

$$\sum_{n=1}^{N} q_n^* - Q = 0$$

$$q_n^* \geq 0$$

$$\beta_n^* \geq 0, \ n = 1, \dots, N$$

$$\beta_n^* q_n^* = 0, \ n = 1, \dots, N. \tag{5.39}$$

With (5.38) we have

$$\frac{\partial L}{\partial q_n} = -\sum_{i=1}^{M_{k_n n}} \frac{\lambda_{k_n n}^{(i)}}{\lambda_{k_n n}^{(i)} q_n^* + N_0} - \alpha^* - \beta_n^* = 0. \tag{5.40}$$

Using (5.39), if $q_n^* > 0$, then $\beta_n^* = 0$. Following (5.40), if $q_n^* > 0$ then it satisfies the following condition:

$$\sum_{i=1}^{M_{k_n n}} \frac{\lambda_{k_n n}^{(i)}}{\lambda_{k_n n}^{(i)} q_n^* + N_0} + \alpha^* = 0, \ n = 1, 2, \dots N. \tag{5.41}$$

As a result, the optimal power distribution across subcarriers is

$$q_n^* = \max(0, q_n), \tag{5.42}$$

where q_n satisfies (5.41) and α^* satisfies $\sum_{n=1}^{N} q_n^* = Q$. Note that this is a "multi-dimensional" water-filling solution, and there is no closed form for q_n^*, α^* and β_n^*. Only numerical results can be evaluated. This finishes the third claim of Theorem 4. ∎

Remark 5 *A special case of the optimal power loading is when $N_t = N_r = 1$. In this case, $M_{kn} = 1$ for all k and all n. The optimal power distribution reduces to the SISO solution in (5.27) which is the power "water-filling" solution [28]:*

$$q_n^* = \left(-\frac{1}{\alpha^*} - \frac{N_0}{\lambda_{k_n n}^{(1)}} \right)^+ ,$$

where $(\cdot)^+ = max(0, \cdot)$ and α satisfies

$$\sum_{n=1}^{N} \left(-\frac{1}{\alpha} - \frac{N_0}{\lambda_{k_n}^{(1)}} \right)^+ = Q.$$

Subcarrier allocation criteria

While Theorem 4 establishes the optimality of OFDMA/MIMO, it does not explicitly reveal how to perform the optimal power and subcarrier allocation. This is because the optimal subcarrier allocation criterion in (5.29) and the optimal power loading criterion in (5.30) cannot be applied separately. To see this, notice that in order to determine the assigned user (k_n) for a specific subcarrier n, the amount of power assigned to this subcarrier (q_n^*) has to be known. On the other hand, to determine the optimal power distribution (q_n^*) across subcarriers, the assigned user (k_n) has to be known. As a result, in order to perform the optimal subcarrier allocation and the optimal power loading, an exhaustive search is needed. To quantify the complexity, note that for each subcarrier there are K possibilities of k_n. Therefore there are $K \cdot K ... \cdot K = K^N$ possibilities of subcarrier allocation over all subcarriers. For each of the K^N outcome of subcarrier allocation, a power loading based on (5.30) is needed to determine which one of these outcomes yields the highest capacity. The cost of the exhaustive search is thus exponential with respect to N and polynomial with respect to K.

To simplify the subcarrier allocation and power loading, we propose two subcarrier allocation criteria, namely, Product-criterion and Sum-criterion, which are independent of the power loading process:

$$k_n^{(P)} = \arg\max_{k} \prod_{i=1}^{M_{kn}} \lambda_{kn}^{(i)}. \tag{5.43}$$

$$k_n^{(S)} = \arg\max_{k} \sum_{i=1}^{M_{kn}} \lambda_{kn}^{(i)}. \tag{5.44}$$

The removal of the power term will allow straightforward subcarrier allocation. As a result, the computation complexity of performing subcarrier allocation and power loading is KN.

To justify such simplifications, notice that in the large SNR region, i.e., $\frac{\lambda_i q_n}{N0} \gg 1$, we obtain the following approximations from (5.29)

$$
\arg\max_k \sum_{i=1}^{M_k} \log\left(1 + \frac{\lambda_k^{(i)} q_n}{N0}\right) = \arg\max_k \log \prod_{i=1}^{M_k}\left(1 + \frac{\lambda_k^{(i)} q_n}{N_0}\right)
$$

$$
\approx \arg\max_k \log \prod_{i=1}^{M_k} \frac{\lambda_k^{(i)} q_n}{N_0}
$$

$$
= \arg\max_k \prod_{i=1}^{M} \lambda_k^{(i)} \quad \text{when } M_1 = ... = M_K = M.
$$

Therefore the Product-criterion tends to be more accurate when the SNR is high. On the other hand, in the small SNR region, i.e., $\frac{\lambda_i q_n}{N0} \ll 1$, using $log(1 + x) = x$, we obtain the following approximations:

$$
\arg\max_k \sum_{i=1}^{M_k} \log\left(1 + \frac{\lambda_k^{(i)} q_n}{N0}\right) \approx \arg\max_k \sum_{i=1}^{M_k} \frac{\lambda_k^{(i)} q_n}{N0}
$$

$$
= \arg\max_k \left(\sum_{i=1}^{M_k} \lambda_k^{(i)}\right) \frac{q_n}{N_0}
$$

$$
= \arg\max_k \sum_{i=1}^{M_k} \lambda_k^{(i)}.
$$

Therefore the Sum-criterion tends to be more accurate when the SNR is low.

After the subcarrier allocation is determined, the optimal power loading can be deployed based on (5.30). With the proposed criteria, the computation complexity to perform subcarrier allocation and power loading is only KN.

We now compare the two suboptimal criteria (5.43) and (5.44) against the optimal criterion using exhaustive search in the downlink of the multiuser multicarrier MIMO system.

Example 23 *Figure 5.18 shows the capacity per subcarrier per unit power versus SNR with $K = N = 8$ and $N_t = N_r = 2$ in the low SNR region. Figure 5.19 shows the same scenario in the high SNR region. As we have predicted, the Product-criterion yields higher capacity than the Sum-criterion in the high SNR region. As a matter of fact, its performance is very close to that of the optimal subcarrier allocation scheme obtained through the exhaustive search. On the other hand, the Sum-criterion yields higher capacity than Product-criterion in the small SNR region. However the capacities obtained using the three criteria are very close in the low SNR region.*

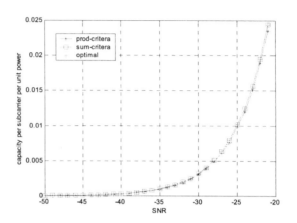

Figure 5.18: Capacity per subcarrier per unit power vs. SNR. $N_t = N_r = 2$; $Q = N$

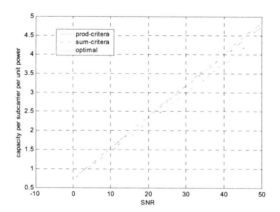

Figure 5.19: Capacity per subcarrier per unit power vs. SNR. $N_t = N_r = 2$; $Q = N$

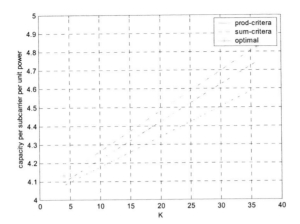

Figure 5.20: Capacity per subcarrier per unit power vs. the number of users. $N_t = N_r = 4$; $N = 8$ and $SNR = 20$dB

Example 24 *Figure 5.20 gives the capacity per subcarrier per unit power versus the number of users when $N = 8$, $SNR = 20$dB and $N_t = N_r = 4$. It is observed that the total capacity increases with the number of users. The capacity increase can be explained by the multiuser diversity gain – as the number of users increases, the probability of a high quality user on a given subcarrier increases too, resulting in more efficient power usage and a larger capacity.*

5.6 Summary

In this chapter, we first overview the two mostly used wireless multiple access control protocols, namely, contention- and non-contention-based protocols. Each protocol is briefly described with examples in practical systems. The emphasis of this chapter is on multiuser diversity in OFDMA networks. We define the multiuser diversity gain as the system performance improvement due to diverse utility values perceived by different users on each radio resource. We prove the optimality of OFDMA in generic downlink multiuser SISO/MIMO models. For practical application, such as the WiMAX OFDMA mode, we present two OFDMA subcarrier and power allocation criteria/algorithms suitable for real-time implementation.

Appendix I: $C_n(\mathbf{p})$ is a convex function

We prove the statement with the following Lemma [27]:

Lemma 3 $f(x)$ *is (strictly) convex if the Hessian of f is positive (definite) semi-definite.*

We now show that the Hessian of $C_n(p_1, ..., p_K)$ is positive definite. Recall that

$$C_n(p_1, ..., p_K) = \sum_{k=1}^{K} \log \left(1 + \frac{p_k}{q_n - p_k + N_0/\|h_k\|^2} \right).$$

Therefore,

$$\frac{\partial^2 C_n}{\partial p_k^2} = \frac{1}{\left(q_n - p_k + N_0/\|h_k\|^2 \right)^2} > 0,$$

$$\frac{\partial^2 C_n}{\partial p_i \partial p_j} = 0, \text{ for } i \neq j.$$

It is obvious that the Hessian of $C_n(p_1, ..., p_K)$ is strictly positive definite. With Lemma 3, we state that $C_n(p_1, ..., p_K)$ is strictly convex.

Appendix II: $C(\mathbf{p})$ is a convex function

We prove the statement with the following Lemma:

Lemma 4 *If $f_1(x)$ and $f_2(x)$ are both convex functions, then $f_1(x) + f_2(x)$ is also a convex function with the domain of dom $f_1 \cap$ dom f_2 [27].*

We now show that the Hessian of $C_n(p_1, ..., p_K)$ is positive definite. Recall that

$$\widehat{c}_k(p_1, ...p_K) \triangleq c_k(g_k) = \log \prod_{i=1}^{M_k} \left(1 + \frac{\lambda_k^{(i)} p_k}{\lambda_k^{(i)}(q_n - p_k) + N_0} \right)$$

$$= \sum_{i=1}^{M_k} \log \left(1 + \frac{\lambda_k^{(i)} p_k}{\lambda_k^{(i)}(q_n - p_k) + N_0} \right)$$

Therefore,

$$\frac{\partial^2 \widehat{c}_k}{\partial p_k^2} = \sum_{i=1}^{M_k} \frac{\left(\lambda_k^{(i)} \right)^2}{\left(\lambda_k^{(i)}(q_n - p_k) + N_0 \right)^2} > 0,$$

$$\frac{\partial^2 \widehat{c}_k}{\partial p_i \partial p_j} = 0, \; i \neq j.$$

It is obvious that the Hessian of $\widehat{c}_k(p_1, ..., p_K)$ is strictly positive definite. Therefore with Lemma 3, we state that $\widehat{c}_k(p_1, ..., p_K)$ is strictly convex. With Lemma 4, we then conclude that $C_n(p_1, ..., p_K) = \sum_{k=1}^{K} c_k(g_k) = \sum_{k=1}^{K} \widehat{c}_k(p_1, ..., p_K)$ is also a convex function.

Bibliography

[1] ETSI Specification: GSM 05.01 (ETS 300 573): Digital cellular telecommunications system (Phase 2); Physical layer on the radio path General description.

[2] IEEE Std 802.11-1997 IEEE Std 802.11-1997 Information Technology-telecommunications And Information exchange Between Systems-Local And Metropolitan Area Networks-specific Requirements-part 11: Wireless Lan Medium Access Control (MAC) And Physical Layer (PHY) Specifications.

[3] IEEE Std 802.15.1 IEEE Standard for Information technology- Telecommunications and information exchange between systems- Local and metropolitan area networks- Specific requirements Part 15.1: Wireless Medium Access Control (MAC) and Physical Layer (PHY) Specifications for Wireless Personal Area Networks (WPANs).

[4] R. Lupas and S. Verdu, "Near-far resistance of multiuser detectors in asynchronous channels,*IEEE Trans. on commun.*, vol. 38, no. 4, April, 1990.

[5] L. Harte, M. Hoenig, D. McLaughlin, R. Kikta, *CDMA IS-95 for Cellular and PCS: Technology, Applications, and Resource Guide*, McGraw-Hill Professional, 1999, ISDN 0070270708.

[6] D. Kivanc, G. Li and H. Liu, "Computationally efficient bandwidth allocation and power control for OFDMA,*IEEE Trans. Wireless Commun.*, vol. 2, Nov. 2003, pp.1150 -1158.

[7] IEEE Standard for Local and Metropolitan Area Networks Part 16: Air Interface for Fixed Broadband Wireless Access Systems.

[8] G. J. Pottie, "System design choices in personal communications,*IEEE Personal Communication*, vol. 2, no. 5, pp. 50 -67, Oct. 1995.

[9] M. Moeneclaey, M. Van Bladel and H. Sari, "Sensitivity of multiple- access techniques to narrow-band interference, *IEEE Trans. on Commun.*, March, 2001.

[10] I. Emre Telatar, "Capacity of Multi-antenna Gaussian Channels,*European Trans. Telecommun.*, vol. 10, pp. 585-595, Nov. 1999.

[11] S. Shakkottai, T. S. Rappaport and P. C. Karlsson, "Cross-layer design for wireless networks,*IEEE Commun. Mag.*, Oct., 2003.

[12] M. Grossglauser and D. Tse, "Mobility increases the capacity of ad-hoc wireless networks,*IEEE/ACM Trans. Networking*, vol. 10, Aug. 2002, pp. 477-486.

[13] H. Liu, *Signal processing applications in CDMA communications*, Boston, Artech House, c2000.

[14] S. M. Alamouti, "A Simple Transmitter Diversity Scheme for Wireless Communications,*IEEE J. Selected Areas in Commun.*, vol. 1, pp. 1451–1458, October 1998.

[15] R. S. Blum, Y. Li, J. H. Winters, and Q. Yan. "Improved space-time coding for MIMO-OFDM wireless communications,*IEEE Commun. Lett.*, 48:1873–1878, November 2001.

[16] H. Bolckei, D. Gesbert, and A.J. Paulraj. "On the capacity of OFDM-based spatial multiplexing systems,*IEEE Trans. Commun.*, 50(2):225–234, Feb. 2002.

[17] A. Goldsmith, S. A. Jafar, N. Jindal and S. Vishwanath, "Capacity limites of MIMO channels,*IEEE J. on Selected Area in Commun.*, vol. 21, pp. 684-702. June 2003.

[18] E. G. Larsson and P. Stoica, *Space-time block coding for wireless communications*, Cambridge University Press, Cambridge, England, 2003.

[19] A. Paulraj, R. Nabar and D. Gore, *Introduction to Space-time wireless communications*, Cambridge University Press, Cambridge, England, 2003.

[20] L. Shao, S. Roy, S. Sandhu, "Rate-one space frequency block codes with maximum diversity gain for MIMO-OFDM,*IEEE Global Telecommun. Conf., 2003,* Volume: 2 , 1-5 Dec. 2003. pp. 809 - 813.

[21] Y. Huang, J. A. Ritecy, "Tight BER bounds for iteratively decoded bit-interleaved space-time coded modulation,*Commun. Lett., IEEE* , Volume: 8 , Issue: 3 , March 2004. Pages:153 - 155.

[22] K. Liu, V. Raghavan, A. M. Sayeed, "Capacity scaling and spectral efficiency in wide-band correlated MIMO channels,*IEEE Trans. Information Theory*, Volume: 49 , Issue: 10 , Oct. 2003. pp. 2504 – 2526.

[23] Z. Wang and G. B. Giannakis; "Outage mutual information of space-time MIMO channels,*IEEE Trans. Information Theory*, Volume: 50 , Issue: 4 , April 2004. pp. 657 – 662.

[24] J. Jang and K. B. Lee, "Transmit power adaptation for multiuser OFDM systems,*IEEE J. Selected Areas in Commun., vol.21, no. 2, Feb. 2003.*

[25] R. Tyrrell Rockafellar, *Convex analysis*, Princeton, NJ., Princeton University Press, 1970.

[26] J. Nocedal and S. J. Wright, *Numerical Optimization*, Springer-Verlag, ISBN 0387-98793-2, New York, 1999.

[27] S. Boyd and L. Vandenberghe, *Convex Optimization*, Cambridge University Press, Cambridge, England, 2004.

[28] T. M. Cover and J. Thomas, *Elements of Information Theory*, Wiley series in Telecommunications, John Wiley & Sons, 1991.

[29] B. Suard, G. Xu, H. Liu and T. Kailath, "Uplink channel capacity of space-division-multiple-access schemes,*IEEE Trans. Information Theory,* vol. 44, no. 4, pp. 1468-1476, July 1998.

Chapter 6

OFDMA Design
Considerations

While Chapter 5 establishes the OFDMA optimality in theory, many practical issues must be addressed before OFDMA can be utilized in real applications. In this chapter, we discuss some PHY and MAC design aspects of practical OFDMA systems. In particular, we introduce a cross-layer concept that benefits the OFDMA system performance by taking terminal mobility into account in OFDMA traffic channel design. The idea of dynamically reconfigurable traffic channel has been an active topic in IEEE 802.16e standardization activities.

6.1 Cross-layer design introduction

Cross-layer design is the concept of joint optimization over multiple layers in system design. In traditional systems, multiple layers are decoupled and layers do not operate interactively. As a result, each layer is designed to cope with the worst case condition, leading to inefficient usage of spectrum and energy. The separation of multiple layers are problematic in practice when handling the integration of mixed services in modern wireless networks. In high-speed data networks, traffic is highly diverse with distinct QoS parameters; channel and environmental conditions may vary dramatically on a short time scale; even the user pattern presents high dynamics – some users are highly mobile, while others may be semi-stationary. These added dynamics require a set of adaptive protocols to cope with these variations. However, the traditional decoupled layer design cannot meet such requirements. For instance, if the MAC layer does not interact with upper layers, it cannot obtain information regarding the type of service and the associated QoS parameters. As a consequence, MAC has no ability to adjust itself to the dynamic traffics.

A simple cross-layer design requires the adaptability in MAC and PHY layer in response to application services. For example, in order to support integrated voice and data services, the MAC layer has to distinguish the type of service and its associated QoS parameters, and map the data service to a set of appropriate physical layer configurations. Delay sensitive services such as teleconferencing can be supported with less robust PHY parameters to yield less delay; while the error-sensitive services should be configured with stronger codes and more stringent PHY parameters to protect the data. Such a process needs the interaction of PHY, MAC and upper layers.

A commonly used cross-layer approach is the Hybrid ARQ (H-ARQ), also called incremental redundancy ARQ. H-ARQ involves the interaction of MAC and PHY layers. In general, the MAC layer transmits several subpackets belonging to one encoded block in the stop-and-wait fashion, based on ACK/NAK fed back from the receiver. H-ARQ has the advantage of coping with the dynamic channels and interference conditions adaptively. H-ARQ brings performance improvement due to the improved SNR gain, time diversity and incremental redundancy by combining the previously received subpacket and the retransmitted packet.

Example 25 *H-ARQ is optional in IEEE 802.16e [1] and may be activated on a per-terminal basis. Each H-ARQ packet is encoded with corresponding PHY specifications and is divided into four subpackets. In downlink communications, the BS transmits one subpacket at a time and waits for the ACK/NAK from the terminal. Due to the redundancy embedded in the packet, the terminal may decode the information correctly without having to receive all four subpackets. In this case, the terminal indicates to the BS with an ACK, otherwise a NAK is sent to the BS to trigger the next subpacket transmission. In order to facilitate the fast feedback of ACK/NAK responses, IEEE 802.16 specifies a dedicated PHY layer ACK/NAK channel in uplink and an H-ARQ ACK message in downlink.*

Efficiently delivering TCP data over wireless links also requires cross-layer design approach. Conventional TCP regards packet loss as an indication of network congestion and, thus, decreases the transmission rate accordingly. However, delivering TCP packets over wireless links has different characteristics from the wired link. Packet loss due to bad channel conditions needs to be addressed differently from network congestion – the channel-quality-related packet loss should not incur the decrease of TCP transmission rate. One approach is to deploy ARQ and coding at a faster time scale than the TCP control loop [2]. Such a scheme hides TCP from the wireless links and the channel perceived by TCP layer is almost invariant. However, this approach loses capacity due to the extra effort put on the lower layers.

Example 26 *A cross-layer approach of handling the TCP over wireless problem is to distinguish the packet loss caused by wireless links and network congestion, then indicate it to the upper layers using an ECN mark bit [3]. It is shown that by explicit loss notification, an improved TCP performance is achieved [2].*

6.2   Mobility-dependent traffic channels

As stated in Chapter 5, an OFDMA system possesses rich multiuser diversity that can be exploited through intelligent resource allocation. Towards this end, the MAC layer must distinguish the user pattern/service model and adapts MAC and PHY to the appropriate configurations. In the ensuing sections, we discuss a new cross-layer design approach, namely, *mobility-dependent traffic channel configuration and allocation*, to allow MAC and PHY layers to adapt to the user pattern and service models. The results presented here allow OFDMA system designers to quantify the trade-offs and arrive at the optimum configuration based on the types of services supported.

More specifically, in this cross-layer design concept, we define two types of services: fixed/portable application and mobile application. The main difference between them resides in whether intelligent resource allocation will be implemented in the MAC layer.

- Fixed/portable application

  Fixed/portable applications are services that are operated in very slow fading environments. In this case, the channel is varying very slowly and the channel estimation is relatively accurate. The channel measurement report/feedback and the channel re-allocation do not need to update very often, thus the incurred signaling overhead is acceptedly low. As a consequence, the MAC layer is equipped with the real-time channel state information without too much overhead. Therefore, it is very reasonable to exploit the OFDMA multiuser diversity through intelligent resource allocation in MAC layer.

- Mobile application

  Mobile applications are services operated in very fast fading environments such as mobile wireless Metropolitan Area Network (MAN). In a fast fading environment, the measurements must be performed and fed back very often in order to track the rapidly varying channels accurately. The channel re-allocation, if present, also needs to be updated quickly to synchronize with the fast fading channels. Thus, resource allocation consumes an intensive signaling overhead. As a result, exploiting multiuser diversity with resource allocation is infeasible in fast fading environments. In this case, the PHY layer must cope the fast fading channels with proper interleaving and spreading to yield sufficient time and frequency diversity.

The cross-layer design concept described below distinguishes the two types of applications and adapts PHY layer with the appropriate channel configurations.

6.2.1   OFDMA traffic channel

Definition 5 *OFDMA traffic channel configuration is defined as the grouping and configuring of OFDM subcarriers into basic resource units – traffic channels*

Theoretically, each OFDM subcarrier can be assigned to a different user. In practice, however, a single-subcarrier based traffic channel is hard to implement and may be too small to provide the basic services.

Example 27 *The system profile OFDMA_ProfP1 in IEEE 802.16e has a channel bandwidth of 1.25 MHz. Using a 2048-point FFT and the lowest modulation and parameters allowed (QPSK and 1/2 code rate), each subcarrier can maximally support $\frac{1.25}{2048} \times 2 \times \frac{1}{2} = 0.61$ Kbps service. The maximum supportable rate of each subcarrier in the example is too small to provide the basic service even for voice services.*

Traffic channel configuration is an important design factor in OFDMA systems. Intuitively speaking, for fixed/mobile applications where the MAC layer can exploit the multiuser diversity, one should increase the multiuser diversity by grouping consecutive subcarriers to enable maximum gain variation over traffic channels. In other words, the traffic channels should be configured so that users' achievable rates have the highest variance (i.e., highest multiuser diversities). A "tight" traffic channel comprising consecutive (near coherent) subcarriers is thus more desirable for fixed applications. On the other hand, for mobile applications in fast fading scenarios where dynamic channel allocation is prohibitive, "spread" traffic channels with low variance in achievable rate are preferred.

6.2.2 System model

Consider an OFDMA system with a total number of N subcarriers and K users. We divide the N subcarriers into L traffic channels, each with M subcarriers: $N = L \times M$. In downlink transmission, information bits from the base station are first encoded by the channel encoder, interleaved and then mapped to modulation symbols. The modulated symbols are loaded onto the traffic channel allocated to the particular user. Depending on the channel conditions, a proper modulation scheme (e.g., QPSK, 64QAM) and the code rate are selected. In some cases, a symbol may be transmitted over multiple subcarriers for frequency diversity.

Generally speaking, the mapping from subcarriers into traffic channels can be arbitrary. In practice, however, regular mapping is often utilized for easy implementations. In regular mapping, the M subcarriers in each traffic channel are further divided into M/M_c clusters, with each cluster having M_c consecutive subcarriers. The distance between clusters within the same traffic channel is fixed for all traffic channels and the spacing between two clusters is $M_d = N/(M/M_c)$ – please see Figure 6.1 for illustrations. Under these definitions, a regular traffic channel configuration can be uniquely defined by $\{(M, M_c)\}$.

Example 28 *In Figure 6.1, the 32 subcarriers are divided into four traffic channels and each traffic channel has 8 subcarriers. The 8 subcarriers in the same traffic channel are further divided into two clusters with each cluster having 4*

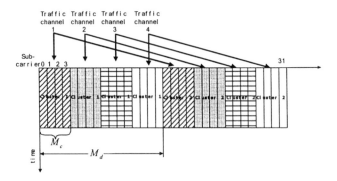

Figure 6.1: Illustration of the traffic channel configuration: $\mathcal{C}=\{(8,4)\}$

consecutive subcarriers. Hence, this configuration is denoted as $\{(8,4)\}$ and the cluster spacing is 16.

Clearly, an $\{(M, M)\}$ configuration has the tightest traffic channel. Its corresponding achievable rate has the highest variance among users. On the other hand, an $\{(M, 1)\}$ configuration has the most spread subcarriers with the highest frequency diversity. In this section, we study how different choices of cluster size M_c and traffic channel size M affect the overall system performance. As will be shown, different choices of M_c and M provide various trade-offs in fixed and mobile applications.

Different traffic channel configurations lead to different correlations among subcarriers in the same traffic channel, consequently, different levels of frequency diversity. In order to study the correlation among subcarriers, we model a broadband channel in a Rayleigh fading environment as Υ sufficiently resolvable uncorrelated paths with normalized (by symbol duration T_s) delays: $\tau_0, \tau_1, ..., \tau_{\Upsilon-1}$ ($\tau_0 = 0$). We further assume that the Υ path gains are independent complex Gaussian random variables:

$$\boldsymbol{\alpha} = [\alpha_0, \alpha_2, ..., \alpha_{\Upsilon-1}]^T,$$

where $\{\alpha_i\}$ have independent real and imaginary parts with zero mean and variance $\sigma_i^2/2$, $i = 0, 1, ...\Upsilon - 1$.

We define the channel (frequency) response vector as

$$\mathbf{h} = [h_0, h_1, ..., h_{N-1}]^T,$$

where

$$h_n = \sum_{i=0}^{\Upsilon-1} \alpha_i e^{-j\frac{2\pi\tau_i n}{N}}, n = 0, 1, ..., N - 1.$$

Denoting $w_N = e^{-j\frac{2\pi}{N}}$, the channel vector can be re-written as:

$$\mathbf{h} = \mathbf{W}\boldsymbol{\alpha},$$

where

$$\mathbf{W}_{N \times \Upsilon} = [(w_N)^{n\tau_i} : n = 0, 1, ..., N - 1; \; i = 0, 1, ..., \Upsilon - 1].$$

It can be easily verified that h_n, $n = 0, 1, ..., N$ are identically distributed but correlated zero mean complex Gaussian variables with equal variance $\sum_{i=0}^{\Upsilon-1} \frac{\sigma_i^2}{2}$ on real and imaginary parts.

Let x_n and y_n be the transmitted and received symbol on the nth subcarrier, respectively. Assuming that the length of the cyclic prefix is greater than the length of the channel response, we have [4]:

$$y_n = h_n x_n + v_n, \; n = 0, 1, ..., N - 1, \tag{6.1}$$

where the AWGN noise v_n has variance N_0. Accordingly, we arrange the transmitted signals, the received signals, the channel responses, and the noise into the vector forms:

$$\begin{aligned}
\mathbf{x} &= [x_0, x_1, ..., x_{N-1}]^T, \\
\mathbf{y} &= [y_0, y_1, ..., y_{N-1}]^T, \\
\mathbf{h} &= [h_0, h_1, ..., h_{N-1}]^T, \\
\mathbf{v} &= [v_0, v_1, ..., v_{N-1}]^T,
\end{aligned}$$

Thus the input-output relation can be modeled as parallel but correlated Gaussian channels:

$$\mathbf{y} = \mathbf{h} \odot \mathbf{x} + \mathbf{v}, \tag{6.2}$$

where $\odot$ denotes element-by-element multiplication.

For the OFDMA system under consideration, let $\Omega_l = \{l_1, ..., l_M\}$ be the configured subcarrier indices of traffic channel l. We define $\mathbf{h}_l = [h_{l_1}, h_{l_2}, ..., h_{l_M}]^T$ as the channel vector corresponding to the lth traffic channel. If needed a superscript k will be added (i.e., $\mathbf{h}_l^k$) to denote the kth user's channel response on the lth traffic channel.

In the following, we derive the optimal channel configuration for fixed/portable applications and mobile applications, respectively.

6.2.3 Channel configuration for fixed applications

For fixed or portable services with static or quasi-static channels, $\mathbf{h}$ can be treated as a constant in a resource allocation period. In this case, centralized traffic channel allocation is assumed to capture the multiuser diversity gain. Users' channel state information is periodically sent back to the BS. The BS then makes the resource allocation decision for users and selects the corresponding adaptive coding and modulation on each traffic channel The resource allocation decision and the ACM scheme are sent back to the users, who decode their information on the allocated channels according to the selected ACM. This is feasible since both the transmitter and the receiver can have accurate channel state information with relatively low signaling overhead. Intuitively, the more diverse the users' channel responses are, the more likely that the BS finds a user with high channel quality on a given traffic channel.

In the following, we shall use the averaged aggregated rate to analyze different configurations. The averaging over all channel realizations will be used as a performance measure. To this end, we derive the statistical characteristics of $\mathbf{h}_l$, based on which we calculate the downlink aggregated rate.

Aggregated rate with channel allocation

In practice, the actual mapping between the vector channel and the achievable rate depends on the available ACM schemes and the BER target. For simplicity, we assume that the average SNR of the subcarriers in a traffic channel is used in ACM selection. As a result, the achievable data rate of the kth user on the lth traffic channel, r_l^k, can be expressed as

$$r_l^k = g(\mathbf{h}_l^k),$$

where $g(\cdot)$ represents the rate-SNR function of the ACM scheme. The average SNR experienced by the kth user on the lth traffic channel, s_l^k, is given as

$$s_l^k = \frac{\delta_l^k}{MN_0},$$

where N_0 is the noise variance and δ_l^k is the channel gain on the lth traffic channel calculated as

$$\delta_l^k = \|\mathbf{h}_l^k\|^2. \tag{6.3}$$

Using ACM, the achievable data rate is then expressed as:

$$\begin{aligned} r_l^k &= M \cdot g(s_l^k) \\ &= M \cdot g\left(\frac{\delta_l^k}{MN_0}\right). \end{aligned}$$

To focus on the effect of channel configuration, we normalize the downlink channel to unit power to remove the propagation loss factor, i.e., $E\{\|h_n\|^2\} = 1$.

Clearly, the downlink aggregated rate is maximized when each traffic channel is allocated to the user with the maximum achievable rate on that particular channel. Then the channel allocation criterion is

$$k_l = \arg\max\{r_l^1, r_l^2, ...r_l^K\}, \tag{6.4}$$

and the maximum rate on the lth traffic channel is given by

$$r_{l,\max} = \max\{r_l^1, r_l^2, ...r_l^K\}. \tag{6.5}$$

Note that the rate-SNR function $g(\cdot)$ is non-decreasing with respect to SNR, hence a traffic channel will always be assigned to the user with the highest SNR, or equivalently, the user with the highest channel gain δ_l^k. Then (6.5) can be re-written as

$$r_{l,\max} = M \cdot g \left(\frac{\delta_{l,\max}}{M N_0} \right),$$

where

$$\delta_{l,\max} = \max\{\delta_l^1, \delta_l^2, ... \delta_l^K\}.$$

The system aggregated rate is then given by

$$r_{sys} = \sum_{l=1}^{L} r_{l,\max} = M \sum_{l=1}^{L} g \left(\frac{\delta_{l,\max}}{M N_0} \right), \tag{6.6}$$

and the normalized averaged aggregated rate (bits/s/Hz) is

$$\bar{r}_{sys} = r_{sys}/N. \tag{6.7}$$

With the aim of designing traffic channels to achieve the highest system aggregated rate, we now attempt to quantify the multiuser diversity gain of a broadband OFDMA channel. In this section, we derive the expected value of (6.7) under different traffic channel configurations by deriving

- the statistics of the channel gain δ_l^k under different traffic channel configurations;

- the statistics of $\delta_{l,max}$ and, consequently, the expected value of the normalized aggregated rate.

Before proceeding, let us assume that all h_n have the same statistics with normalized mean value $(E\{\|h_n\|^2\} = 1)$. Thus the covariance matrix $\mathbf{R}_{\mathbf{h}_l^k \mathbf{h}_l^k}$ is the same for all k and all l. As a consequence, all δ_l^k have the same probability pdf $f_\delta(x)$ and they are expressed in the following Proposition. (We drop some of the superscripts and subscripts for convenience.)

Proposition 8 δ_l^k *is a random variable with characteristic function* $\psi_\delta(jw)$, *pdf* $f_\delta(x)$ *and CDF* $F_\delta(x)$ *expressed as follows, assuming that all the eigenvalues of* $\mathbf{R}_{\mathbf{h}_l \mathbf{h}_l}$ *are distinct,*

$$\psi_\delta(jw) = \prod_{j=1}^{M_1} \frac{1}{1 - jw\lambda_j}, \tag{6.8}$$

$$f_\delta(x) = \sum_{j=1}^{M_1} \frac{\beta_j}{\lambda_j} e^{-x/\lambda_j}, \ x \geq 0, \tag{6.9}$$

$$F_\delta(x) = \sum_{j=1}^{M_1} \beta_j (1 - e^{-x/\lambda_j}), \ x \geq 0, \tag{6.10}$$

where

$$\beta_j = \prod_{i=1, i\neq j}^{M_1} \frac{\lambda_j}{\lambda_j - \lambda_i},$$

M_1 *is the rank of* $\mathbf{R_{h_l h_l}}$, $\lambda_1, ..., \lambda_{M_1}$ *are the* M_1 *non-zero distinct eigenvalues of* $\mathbf{R_{h_l h_l}}$ *and*

$$\mathbf{R_{h_l h_l}} = E\left\{ \mathbf{h}_l \mathbf{h}_l^H \right\}.$$

In the case of $M_1 = 1$,

$$f_\delta(x) = \frac{1}{\lambda_1} e^{-x/\lambda_1}.$$

Proof. It is easy to see that $\mathbf{R_{h_l h_l}}$ is a Hermitian matrix which is unitarily diagonalizable and its eigenvalues are real [5, Theorem 2.5.6]. Therefore the correlation matrix $\mathbf{R_{h_l h_l}}$ can be represented as

$$\mathbf{R_{h_l h_l}} = \mathbf{U}^H \mathbf{\Lambda} \mathbf{U},$$

where $\mathbf{U}$ is an orthonormal matrix and $\mathbf{\Lambda} = diag(\lambda_1, ..., \lambda_{M_1}, 0, ..., 0)$ with λ_i being the ith eigenvalue of $\mathbf{R_{h_l h_l}}$. When $\mathbf{R_{h_l h_l}}$ is full rank, $M_1 = M$. Denote

$$\widehat{\mathbf{h}}_l = \mathbf{\Lambda}_{M_1}^{-1/2} \mathbf{U}_{M_1} \mathbf{h}_l,$$

where

$$\mathbf{\Lambda_{M_1}} = diag\{\lambda_1^{-1/2}, ..., \lambda_{M_1}^{-1/2}\},$$

and $\mathbf{U}_{M_1}$ contains the first M_1 columns of $\mathbf{U}$. It is easy to show that $\widehat{\mathbf{h}}_l$ are independent by verifying

$$E\{\widehat{\mathbf{h}}_l \widehat{\mathbf{h}}_l^H\} = \mathbf{I}.$$

Therefore, (6.3) can be re-written as

$$\|\delta_l\|^2 = \|\mathbf{h}_l\|^2 = \widehat{\mathbf{h}}_l^H \mathbf{\Lambda}_{M_1} \widehat{\mathbf{h}}_l.$$

Thus δ_l is the sum of M_1 statistically independent random variables each of which satisfies chi-square distribution with two degrees of freedom. Hence the characteristic function of δ_l is

$$\psi_{\delta_l}(jw) = \prod_{j=1}^{M_1} \frac{1}{1 - jw\lambda_j},$$

and the pdf is obtained from it:

$$f_{\delta_l}(x) = \sum_{j=1}^{M_1} \frac{\beta_j}{\lambda_j} e^{-x/\lambda_m},$$

where

$$\beta_j = \prod_{i=1, i \neq j}^{M_1} \frac{\lambda_j}{\lambda_j - \lambda_i}.$$

∎

Next, we derive the pdf of $\delta_{l,max}$ in order to calculate the expectation of (6.7). Since channels from different users are independent, it can be shown through straightforward manipulations that the pdf and CDF of $\delta_{l,\max}$ are, respectively,

$$F_{\delta_{l,\max}}(x) = F_\delta(x)^K, \tag{6.11}$$

$$f_{\delta_{l,\max}}(x) = K F_\delta(x)^{K-1} f_\delta(x). \tag{6.12}$$

Clearly, all $\delta_{l,max}$, $l = 1, ..., L$, have the same pdf, so we drop the subscript l to simplify the notation.

From (6.9) to (6.12), we arrive at the pdf of δ_{max}

$$f_{\delta_{max}}(x) = K \left[\sum_{j=1}^{M_1} \beta_j (1 - e^{-x/\lambda_j}) \right]^{K-1} \sum_{j=1}^{M_1} \frac{\beta_j}{\lambda_j} e^{-x/\lambda_j}. \tag{6.13}$$

With the pdf of δ_{max} at hand and noting that $NL/M = 1$, the average normalized aggregated rate is obtained from (6.6) and (6.7) as

$$E\left[\bar{r}_{sys}\right] = E\left[g\left(\frac{\delta_{max}}{M\sigma^2}\right) \right]. \tag{6.14}$$

Example 29 *Figure 6.2 gives an example of the CDF of δ_{max} under five different traffic channel configurations. It is seen that the traffic channel configuration C_5, which corresponds to the one with the tightest cluster has the highest δ_{max} while C_1, which corresponds to the one with the highest frequency diversity, has the lowest δ_{max}. δ_{max} decreases significantly with the increase of frequency diversity, indicating a critical trade-off in OFDMA traffic channel design. Such results are explained by the enriched multiuser diversity due to higher variations in achievable rate in C_5 and less multiuser diversity associated with C_1. In fixed applications, since we can utilize multiuser diversity with intelligent channel allocation, multiuser diversity has more impact on system performance.*

Aggregated rate upper bound

The aggregated rate derived in (6.14) assumes equal power allocation across all subcarriers. By allowing (i) transmitter power loading and (ii) each subcarrier to be assigned to a different user (i.e., no traffic channel configuration), one can expect an additional increase in the aggregated rate, which in turn provides an upper bound for OFDMA. We summarize the upper bound statement as follows.

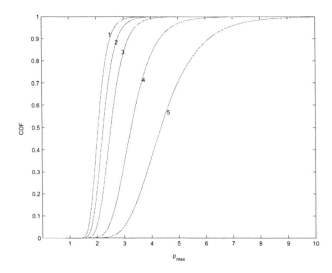

Figure 6.2: CDF of p_{max}. $K = 64$. Traffic channel configurations are: $\mathcal{C}_1=\{(16, 1)\}$, $\mathcal{C}_2=\{(16, 2)\}$, $\mathcal{C}_3=\{(16, 4)\}$, $\mathcal{C}_4=\{(16, 8)\}$, $\mathcal{C}_5=\{(16, 16)\}$

Proposition 9 *The aggregated rate given in (6.14) is upper bounded by the optimal solution to the following problem:*

$$\max_{\mathbf{p}} r'_{sys} = \frac{1}{N} \sum_{n=1}^{N} g\left(p_n \cdot SNR_{n,\max}\right), \qquad (6.15)$$

$$s.t. \sum_{n=1}^{N} p_n = Q,$$

where Q is the total power across all subcarriers. $\mathbf{p} = (p_1, p_2, ..., p_N)$ *are the powers allocated to the subcarriers and $SNR_{n,\max}$ is the highest SNR of subcarrier n over all users.*

Notice that we have solved this problem in Theorem 2, assuming $g(\cdot) = \log(1 + \cdot)$, and the solution to (6.15) is

$$P_n = \left(-\frac{1}{\alpha} - \frac{1}{SNR_{n,\max}}\right)^{+}, \qquad (6.16)$$

where α satisfies

$$\sum_{n=1}^{N} \left(-\frac{1}{\alpha} - \frac{1}{SNR_{n,\max}}\right)^{+} = Q, \qquad (6.17)$$

and $(\cdot)^{+}$ represents $max(\cdot, 0)$. Here we set $Q = N$ in order to make a fair comparison between the solution of (6.15) and the result of (6.14). In (6.14), unit power is allocated to each subcarrier.

The water-filling solution expressed in (6.16) indicates that in order to maximize the system aggregated rate, more power should be allocated to the subcarrier that has better channel quality, i.e., high $SNR_{n,\max}$ leads to high p_n and low $SNR_{n,\max}$ leads to low p_n. $-\frac{1}{\alpha}$ is regarded as the "water level" since the sum of positive p_n and $SNR_{n,\max}$ is a fixed value. Interestingly, we will show that transmitter power control only provides a marginal gain over the result obtained with equal-powered traffic channels configured with the highest multiuser diversity. The advantage of OFDMA channel allocation without power control is evident in terms of its optimality and simplicity.

Aggregated rate lower bound

So far, our design objective has been maximizing the system aggregated rate in (6.14). Under this criterion, users with good channel qualities are likely to get more channels than users in deep fading. In this subsection, we study a progressive channel allocation (PCA) scheme which accounts for the fairness among users. It also serves as a rate lower bound for the centralized channel allocation scheme in (6.4).

We assume that there are as many users as traffic channels. A user requesting multiple traffic channels will be treated as multiple users with identical channel profiles. The PCA scheme introduced here guarantees the channel level fairness with the following simple rules:

- The scheme performs progressively, i.e., one by one, traffic channel allocation;

- Once a user receives one traffic channel, it cannot be assigned any additional channels;

- Each traffic channel is assigned to the best remaining user.

The aggregated rate reached by the PCA channel allocation scheme is obviously lower than the aggregated rate derived in (6.14). In other words, it serves as a lower bound to (6.14), which will be verified shortly. Without loss of generality, let us further assume that channels are assigned in the reversed order of their indices, i.e., channel L is assigned to the best user out the L users; channel $L - 1$ is assigned to the best user out of the remaining $L - 1$ users, so on and so forth.

Denote $\delta^{l,\max}$ as the highest channel gain out of the available users when assigning traffic channel l, it is easy to see that

$$F_{\delta^{l,\max}}(x) = \Pr(\delta^{l,\max} < x) \tag{6.18}$$

$$= \prod_{i=1}^{l} F_{\delta_i^l}(\delta_i^l < x) \tag{6.19}$$

$$= F_\delta(x)^l, \tag{6.20}$$

and

$$f_{\delta^{l,\max}}(x) = lF_\delta(x)^{l-1}f_\delta(x),$$

where δ_i^l is the channel gain of the ith available user when assigning traffic channel l. $f_\delta(x)/F_\delta(x)$ stands for the pdf/CDF of δ_i^l for all l and i which are expressed in (6.9) and (6.10).

Let $r^{l,\max} = M \cdot g(\frac{\delta^{l,\max}}{MN_0})$, indicating the aggregated rate on traffic channel l using PCA scheme, the normalized data rate is then:

$$\widetilde{r}_{sys} = \frac{1}{N}\sum_{l=1}^{L} M \cdot g\left(\frac{\delta^{l,\max}}{M \cdot N_0}\right), \tag{6.21}$$

and

$$E\{\widetilde{r}_{sys}\} = \frac{M}{N}\sum_{l=1}^{L} E\left\{g\left(\frac{\delta^{l,\max}}{M \cdot N_0}\right)\right\}. \tag{6.22}$$

The following proposition shows that the aggregated rate reached by the PCA channel allocation scheme serves as a lower bound to (6.14).

Proposition 10 *The PCA scheme provides a lower bound to (6.14), i.e.,*

$$E\{\widetilde{r}_{sys}\} \le E\{\overline{r}_{sys}\}. \tag{6.23}$$

Proof. Since $l \le K = L$ (in PCA, the number of users is the same as the number of traffic channels) and $F_\delta(x) \le 1$, we obtain

$$F_{\delta^{l,\max}} \ge F_{\delta_{l,\max}}. \tag{6.24}$$

Since

$$E\left\{g\left(\frac{\delta^{l,\max}}{M \cdot N_0}\right)\right\} = \int_0^\infty g\left(\frac{x}{M \cdot N_0}\right) f_{\delta^{l,\max}}(x)dx$$

$$= g\left(\frac{x}{M \cdot N_0}\right) F_{\delta^{l,\max}}(x)\Big|_0^\infty - \int g'\left(\frac{x}{M \cdot N_0}\right) F_{\delta^{l,\max}}(x)dx \tag{6.25}$$

$$= G - \int g'\left(\frac{x}{M \cdot N_0}\right) F_{\delta^{l,\max}}(x)dx$$

$$\le G - \int g'\left(\frac{x}{M \cdot N_0}\right) F_{\delta_{l,\max}}(x)dx \tag{6.26}$$

$$= E\left\{g\left(\frac{\delta_{l,\max}}{M \cdot N_0}\right)\right\},$$

where $G = g(\infty)$ represents the system capacity or the highest ACM scheme allowed. The inequality (6.26) follows from (6.24) and the assumption that $g(x)$ is a non-decreasing continuous function. From equations (6.14), (6.22) and (6.26), we conclude that $E\{\widetilde{r}_{sys}\} \le E\{\overline{r}_{sys}\}$. ∎

(6.23) asserts that the PCA scheme serves as a lower bound to (6.14). The gap of the aggregated rate of PCA compared to (6.14) is due to the loss of multiuser diversity since fewer users are available for each assigning traffic channel in PCA. At the same time, the PCA scheme does provide a simple approach for practical systems to account for user fairness.

Performance analysis for fixed/portable applications

We now examine the impact of channel configuration for fixed/portable services. The ACM used in the simulations is shown in Figure 6.3. It is obtained by using the 64-state convolution code of various rates (1/2, 2/3, 4/5, and 8/7) combined with different modulation schemes (QPSK, 16QAM, and 64QAM) [6]. A BER of 2×10^{-4} is assumed. Typical Urban (TU, non-hilly) power delay profile, defined in COST207 [7], is used in the simulations.

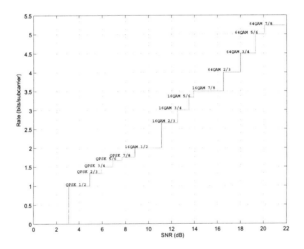

Figure 6.3: Adaptive coding/modulation scheme used in the simulation

Example 30 *Figure 6.4 compares the averaged aggregated rate under different traffic channel configurations with the upper bound using (6.14) and (6.15). The OFDMA system has a bandwidth of 8 MHz, $N = 1024$ subcarriers, and $M = 16$ subcarriers in each traffic channel.*

As expected, C_5 has the highest system aggregated rate among all channel configurations. At $E_s/N_0 = 10dB$, the aggregated rate of C_5 is about 25% higher than that of C_1 (maximum frequency diversity, least multiuser diversity). The 3-4 dB performance gap between C_1 and C_5 essentially quantifies the optimization gain of OFDMA traffic channel design.

Comparing with the upper bound, the aggregated rate corresponding to configuration C_5 ($M_c = M$, i.e., consecutive subcarriers with the maximum multiuser

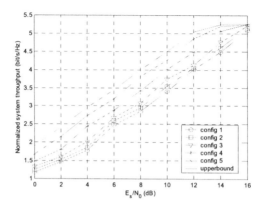

Figure 6.4: Averaged aggregated rate vs. SNR. K=64, TU channel and the traffic channel configurations are: $\mathcal{C}_1=\{(16,\ 1)\}$, $\mathcal{C}_2=\{(16,\ 2)\}$, $\mathcal{C}_3=\{(16,\ 4)\}$, $\mathcal{C}_4=\{(16,\ 8)\}$, $\mathcal{C}_5=\{(16,\ 16)\}$

diversity) is only less than 1 dB below. The small gap indicates an insignificant gain from transmitter side power allocation. In light of the trivial difference and complicated implementation, transmitter side power loading does not seem to be an attractive choice in practice.

Note that the data rate curves flat out at high E_s/N_0 range due to the limit of the ACM scheme. The highest ACM scheme used here is 5.25 bits/symbol corresponding to 64QAM with 7/8 coding.

Example 31 *Figure 6.5 compares the aggregated rates of $\mathcal{C}_5$ and $\mathcal{C}_1$ with their corresponding lower bounds using the PCA channel allocation scheme. By matching $\mathcal{C}_5$ with the PCA scheme, the aggregated rate still outperforms that of $\mathcal{C}_1$ with optimal channel allocation. The degradation due to the use of PCA is roughly 1-2 dB. However, if the traffic channel size (M) is smaller, and, therefore, the number of traffic channels (L) is larger, the degradation due to the use of PCA is not significant as shown in Figure 6.6. The degradation due to PCA is much smaller in configuration $\{(4,\ 4)\}$ than that of $\{(32,\ 32)\}$.*

Based on the above observations, we conclude the following for fixed/portable applications:

- Clusters should be grouped as tight as possible to enable higher multiuser diversity, leading to higher aggregated rate;

- Small traffic channel sizes are preferred – the multiuser diversity gain can be readily captured with the simple PCA scheme.

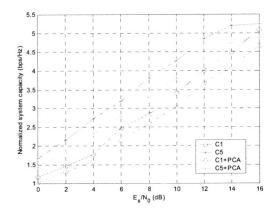

Figure 6.5: Normalized system capacity vs. SNR. K=64, TU channel and traffic channel coinfigurations are: $C_1=\{(16, 1)\}$, $C_5=\{(16, 16)\}$

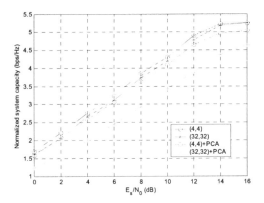

Figure 6.6: Normalized system capacity vs. SNR under different traffic sizes, K=64, TU channel

6.2.4 Channel configuration for mobile applications

While the results for fixed application favor the traffic channel configured with higher multiuser diversity, the situation is intuitively opposite for mobile applications. For mobile applications, we assume that the channel is varying in such a way that it is impractical for the base station/access point to perform meaningful channel allocation. The rapidly varying channel also makes ACM less feasible due to the intensive overhead. The design focus reduces to the achievable rate of a randomly selected user. Here, we study the capacity for two types of services: delay-nonsensitive services and delay-constrained services, respectively. Given that the channel gain is a random process, we resort to the ergodic capacity and the outage capacity [8] [9] to analyze the configuration impact on the system performance. For convenience, we also assume unit power on each subcarrier.

Ergodic case

The basic assumption here is that the fading process is ergodic, i.e., coding and interleaving are performed across OFDM symbols and that the number of OFDM blocks spanned by a code word approaches infinity [10]. In such cases, the ergodic capacity is given as [8]

$$C = \frac{1}{N} \sum_{n=0}^{N-1} E_{h_n}\{I_n\} = \frac{1}{N} \sum_{n=0}^{N-1} E_{h_n} \left\{ \log_2 \left(1 + \frac{\|h_n\|^2}{N_0} \right) \right\}, \qquad (6.27)$$

where I_n is the mutual information on subcarrier n. The ergodic capacity is usually used to determine the maximum achievable long-term rate averaged over all fading realizations. In practice, it is most suitably used to study delay-nonsensitive services such as electronic mails.

Arranging the subcarrier indices in (6.27) according to traffic channels, we obtain

$$C = \frac{1}{L} \sum_{l=1}^{L} C_l,$$

where C_l is the capacity of traffic channel l and is expressed as:

$$C_l = \frac{1}{M} E \left\{ \sum_{m=1}^{M} I_{l_m} \right\} = \frac{1}{M} \sum_{m=1}^{M} E_{h_{l_m}} \left\{ \log_2 \left(1 + \frac{\|h_{l_m}\|^2}{N_0} \right) \right\}, \qquad (6.28)$$

where l_m represents the mth subcarrier in the lth traffic channel.

Clearly I_n depends only on the distribution of h_n. Since $\{h_n\}$ have the same distribution, the frequency correlation among h_n has no effect on either the system capacity C or the capacity of an individual traffic channel C_l. Thus, we have the following proposition:

Proposition 11 *Traffic channel configuration has no effect on the ergodic capacity.*

This result can be explained as follows. When the coding and interleaving length approaches infinity, the received de-interleaved bits are uncorrelated regardless of the correlation in individual OFDM blocks. Therefore, the capacity is the same as that of uncorrelated channels. Consequently, the traffic channel configuration (or correlation among subcarriers) has no effect on the ergodic capacity.

Although traffic channel configuration is not a factor for ergodic capacity, it does contribute to outage capacity in the non-ergodic case, which pertains to most practical scenarios.

Non-ergodic case

Since ergodic capacity only relates to the long-term average rate, it is not a suitable performance measure for delay-constrained services such as voice transmission. In such cases, coding only spans a finite number of OFDM blocks. Shannon capacity does not exist since the mutual information is a random variable dependent on **h**. The concept of outage capacity is invoked [9]. It is also referred to as ε-capacity in [8], which is the capacity guaranteed for $(100-\varepsilon)\%$ of channel realizations. Please refer to [11]-[13] and related references for more details on outage capacity. We will use the following definitions:

Definition 6 *Outage probability for a given rate r, $P_{out}(r)$, is defined as the probability that the mutual information I falls below r: $P_{out}(r) = P(I < r)$.*

Definition 7 *Outage capacity, $r(\varepsilon)$, is the largest r such that the outage probability for this r is less than a given probability ε, i.e., $r(\varepsilon) = \sup\limits_{\{r:\ P_{out}(r)<\varepsilon\}} r$.*

The definition of $P_{out}(r)$ and $r(\varepsilon)$ can be explained using the distribution of I. Denote F_I as the CDF of the mutual information I, then $P_{out}(r) = F_I(r)$, and $r(\varepsilon) = \sup\limits_{\{r:\ F_I(r)<\varepsilon\}} r$.

For simplicity, we assume that independent coding/decoding is performed on each traffic channel. In this case, traffic channel configuration only affects the correlation among the subcarriers within traffic channels. The following proposition states that under the regular configuration, all traffic channels have the same performance.

Proposition 12 *Let $I_l = \frac{1}{M} \sum\limits_{m=1}^{M} \log\left(1 + \frac{\|h_{l_m}\|^2}{N_0}\right)$ be the maximum mutual information of traffic channel l, then $\{I_l : l = 1, 2, ..., L\}$ have the same distribution for all l.*

Proof. Denote $l^{(1)}$ and $l^{(2)}$ as the index of two traffic channels, respectively. Further denote $l^{(1)}_{n_1}$ and $l^{(1)}_{n_2}$ as the index of the n_1^{th} and n_2^{th} subcarriers in traffic channel $l^{(1)}$, respectively. Denote $l^{(2)}_{n_1}$ and $l^{(2)}_{n_2}$ as the index of the n_1^{th} and n_2^{th} subcarriers in traffic channel $l^{(2)}$, respectively. It is easy to see that the distances between the n_1^{th} and n_2^{th} subcarrier in traffic channel $l^{(1)}$ and $l^{(2)}$ are the same, i.e.,

$$l_{n_1}^{(1)} - l_{n_2}^{(1)} = l_{n_1}^{(2)} - l_{n_2}^{(2)}. \tag{6.29}$$

For the claim, it suffices to prove that the covariance matrices of traffic channel $l^{(1)}$ and $l^{(2)}$ are the same, i.e.,

$$E[\mathbf{h}_{l^{(1)}} \mathbf{h}_{l^{(1)}}^H] = E[\mathbf{h}_{l^{(2)}} \mathbf{h}_{l^{(2)}}^H].$$

It is seen that

$$E\{h_{l_{n_1}^{(1)}} h_{l_{n_2}^{(1)}}^H\} = E\left\{ \sum_{i=0}^{\Upsilon-1} \alpha_i e^{-j\frac{2\pi\tau_i l_{n_1}^{(1)}}{N}} \left(\sum_{i=0}^{\Upsilon-1} \alpha_i e^{-j\frac{2\pi\tau_i l_{n_2}^{(1)}}{N}} \right)^H \right\}$$

$$= E\left\{ \sum_{i=0}^{\Upsilon-1} \|\alpha_i\|^2 e^{-j\frac{2\pi\tau_i (l_{n_1}^{(1)} - l_{n_2}^{(1)})}{N}} \right\}.$$

With (6.29), we claim that $E\{h_{l_{n_1}^{(1)}} h_{l_{n_2}^{(1)}}^H\} = E\{h_{l_{n_1}^{(2)}} h_{l_{n_2}^{(2)}}^H\}$ and accordingly,

$$E[\mathbf{h}_{l^{(1)}} \mathbf{h}_{l^{(1)}}^H] = E[\mathbf{h}_{l^{(2)}} \mathbf{h}_{l^{(2)}}^H].$$

∎

The above proposition allows us to reduce our study to the configuration of a single traffic channel. Let us pick traffic channel l and re-label its subcarrier indices from 1 to M :

$$I_l = \frac{1}{M} \sum_{m=1}^{M} \log_2 \left(1 + \frac{\|h_m\|^2}{N_0} \right) \tag{6.30}$$

$$= \frac{1}{M} \log_2 \prod_{m=1}^{M} \left(1 + \frac{\|h_m\|^2}{N_0} \right). \tag{6.31}$$

When the SNR is small, we can approximate I_l as

$$I_l \approx \frac{1}{M} \log_2 \left(1 + \frac{\sum_{m=1}^{M} \|h_m\|^2}{N_0} \right) \tag{6.32}$$

$$= \frac{1}{M} \log_2 \left(1 + \frac{\|\mathbf{h}_l\|^2}{N_0} \right), \tag{6.33}$$

where $\mathbf{h}_l = (h_1, h_2, ... h_M)^T$ are complex Gaussian random variables with covariance matrix $\mathbf{R}_{\mathbf{h}_l \mathbf{h}_l}$. With the aid of (6.3) and (6.10), the CDF of I_l can be calculated through the CDF of $\|\mathbf{h}_l\|^2$ as:

$$P_{out}(r) \quad = \quad F_p(N_0 \left(2^{Mr} - 1\right)) \tag{6.34}$$

$$= \quad \sum_{m=1}^{M_1} \beta_m \left(1 - 2^{-N_0(e^{Mr} - 1)/\lambda_m}\right). \tag{6.35}$$

In an extreme case when the elements of $\mathbf{h}_l$ are mutually independent, i.e., $\lambda_1 = \ldots = \lambda_M = \lambda$, $\|\mathbf{h}_l\|^2$ is a chi-square distribution with $2M$ degree of freedoms and mean $M\lambda$. The outage probability can be calculated as

$$P_{out}(r) = 1 - e^{-x/2\lambda} \sum_{m=0}^{M-1} \frac{1}{m!} \left(\frac{N_0(e^{Mr} - 1)}{2\lambda}\right)^m.$$

When the SNR is large, the following approximation can be used:

$$I_l \approx \frac{1}{M} \sum_{m=1}^{M} \log_2 \left(\frac{\|h_m\|^2}{N_0}\right).$$

In this case, the distribution of I_l is difficult to obtain. In order to make the problem tractable, we approximate $log_2(x)$ as $ax + b$, and the values of a and b depend on the region of SNR. As a result,

$$I_l \approx \frac{1}{M} a \sum_{m=1}^{M} \frac{\|h_m\|^2}{N_0} + b,$$

Using the similar argument for the small SNR region, we obtain the outage probability for the large SNR case as follows:

$$P_{out}(r) = F_\delta \left(\frac{N_0 M(r - b)}{a}\right). \tag{6.36}$$

Performance analysis for mobile applications

We now examine the effect of traffic channel configuration in fast fading channels by studying simulation results. Same simulation parameters are used as that of the fixed applications.

Example 32 *Figure 6.7 gives the CDF of I_l at different SNR levels. It is seen that for the outage probability range of practical interest (e.g., from 0 to 0.4), the outage capacity is the highest for configuration C_1 with the largest frequency diversity, and is the lowest for configuration C_5 with the least frequency diversity. However, if very high outage can be tolerated, the situation is the reverse: C_5 yields the best performance and C_1 is the worst. The observation indicates that in delay sensitive applications, sufficient frequency diversity is the key to system capacity, unless the outage probability permitted by certain applications is extremely high.*

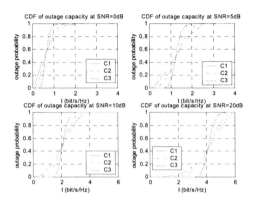

Figure 6.7: CDF of I_l. Three configurations are compared: $\mathcal{C}_1=\{(16, 1)\}$, $\mathcal{C}_2=\{(16, 8)\}$, $\mathcal{C}_3=\{(16, 16)\}$

Example 33 *Figure 6.8 shows outage capacity versus SNR at different outage requirements. Figure 6.9 shows outage capacity versus different outage requirements at SNR=10dB. The two figures only show the situation when outage probability is reasonable (<0.4). As outage probability increases, the differences in system performance corresponding to different configurations diminish. This result indicates that at a modest outage probability, the traffic channel configuration plays an insignificant role in the system performance.*

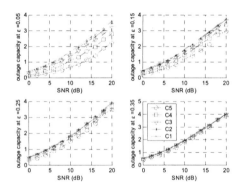

Figure 6.8: Outage capacity vs. SNR. Traffic channel configurations are: $\mathcal{C}_1=\{(16, 1)\}$, $\mathcal{C}_2=\{(16, 2)\}$, $\mathcal{C}_3=\{(16, 4)\}$, $\mathcal{C}_4=\{(16, 8)\}$, $\mathcal{C}_5=\{(16, 16)\}$

Example 34 *Figure 6.10 shows the effect of traffic channel size with configuration $\{(M,1)\}$. Larger traffic channels with more subcarriers yield better per-*

formance. This is due to the fact that the increase of traffic channel size brings more frequency diversity under configuration $\{(M, 1)\}$.

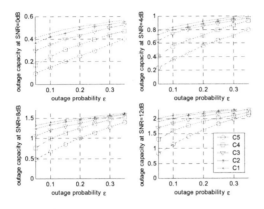

Figure 6.9: Outage capacity vs. outage probability. Traffic channel configurations are: $C_1=\{(16,\ 1)\}$, $C_2=\{(16,\ 2)\}$, $C_3=\{(16,\ 4)\}$, $C_4=\{(16,\ 8)\}$, $C_5=\{(16,\ 16)\}$

Based on the above observations, we conclude the follows for mobile applications:

- Larger traffic channels are preferred to provide better frequency diversity and, thus, higher outage capacity;

- For applications with small outage probability requirements, the clusters should be distributed to enable higher frequency diversity; the opposite is true for applications that can tolerate high outage probabilities.

6.3 IEEE 802.16e traffic channels

For systems with mixed applications, the ideal platform would be the one that supports both "small and tight" traffic channels with maximum multiuser diversity and "large and loose" traffic channels with maximum frequency diversity. The partition can be optimized based the ratio of low mobility and high mobility users. On the other hand, multiple traffic channel configurations in one system may complicate the overall system architecture and resource management. In this case, the optimum trade-off can be calculated based on the results derived in this chapter. Depending on the design objectives (e.g., overall capacity, peak performance etc.), other factors that need to be considered include the ratio between high mobility and low mobility users, ACM schemes and spatial diversity techniques that can enhance the frequency diversity in a fast fading environment.

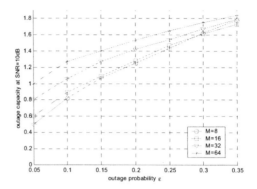

Figure 6.10: Outage capacity vs. outage probability with different traffic sizes. Four configurations are compared: $C_1=\{(16, 1)\}$, $C_2=\{(16, 2)\}$, $C_3=\{(16, 4)\}$, $C_4=\{(16, 8)\}$, $C_5=\{(16, 16)\}$

The IEEE 802.16e has an OFDMA mode with many advanced PHY and MAC features such as traffic channel configuration (termed *subchannel*) discussed previously, zone switching between diversity and adaptive modulation and coding mode, MIMO and advanced antenna system (AAS). The IEEE 802.16e supports variable bandwidth sizes ranging from 1.25 MHz to 20 MHz. The notion of *scalable OFDMA* was introduced in 802.16e to adjust the system parameters to the bandwidth sizes. Specifically, FFT size scales with different bandwidth sizes to fix the subcarriers spacing. The subcarrier spacing design is based on Doppler spread, inter-channel interference (ICI), delay spread and coherence bandwidth etc. [14]. This scalable design has the following advantages:

- Easy implementation and low cost as the subcarrier spacing is a fixed system parameter;

- Maintaining a flat fading on each subcarrier regardless of bandwidth sizes. Such achieves a better performance compared to a fixed FFT size but variable bandwidth OFDMA [17];

- Supporting variable frame sizes ranging from 2ms/19 OFDM symbols to 20ms/198 OFDM symbols. The supportable frame sizes are shown in Table 6.1. Variable frame sizes are flexible to address different types of application and traffic requirement. In addition, variable frame sizes pose a lower bound on the number of OFDM symbols per frame. Such avoids the scenarios with very few OFDM symbols per frame which has a poor spectrum utilization due to high overhead.

The key system parameters of IEEE 802.16e scalable OFDMA are summarized as follows:

Frame sizes (ms)	Frame sizes (OFDMA symbols)
2	19
2.5	24
4	39
5	49
8	79
10	99
12.5	124
20	198

Table 6.1: Scalable OFDMA frame sizes

- Subcarrier spacing is fixed to 11.16 KHz;

- OFDM symbol duration is fixed to 100.8 μs;

- FFT size scales with the bandwidth size;

- The number of traffic channel (termed subchannel) scales with bandwidth size.

Interested readers are referred to the book Appendix for a detailed description.

Example 35 *IEEE 802.16e defines three basic types of subchannel configuration, namely, Fully Used Subchannelization (FUSC), Partially Used Subchannelization (PUSC) and Advanced Modulation and Coding Subchannel (AMC) configuration. FUSC/PUSC bears the same design principles described previously for mobile users, whereas AMC is similar to the traffic channel for fixed/portable users.*

FUSC corresponds to the "loose" and distributed channel configuration that provides high frequency diversity and low multiuser diversity.Figure 6.11 illustrates the 1024-FFT, 10 MHz bandwidth OFDMA FUSC subchannel configuration in 802.16e. Constructively, the bandwidth is divided into 48 groups with each group comprising 16 subcarriers. The system contains 16 subchannels with each subchannel comprising 48 subcarriers. The 48 subcarriers in the same subchannel are taken from the separate 48 groups. The exact configuration of allocating the subcarriers into subchannels is according to a pre-defined DL permutation formula.

AMC subchannel configuration corresponds to the "tight" and adjacent channel configuration discussed for fixed applications. AMC configuration provides low frequency diversity and high multiuser diversity. In AMC configuration, the subcarriers are partitioned into bands of consecutive subcarriers. Each band contains 4 bins, which is a collection of 8 data subcarriers plus one pilot subcarrier. The AMC subchannels can be dynamically allocated for multiuser exploitation. Please see the book Appendix for more details on subchannel configuration in IEEE 802.16e.

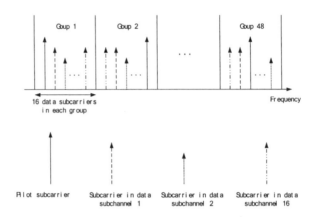

Figure 6.11: 802.16e FUSC channel configuration

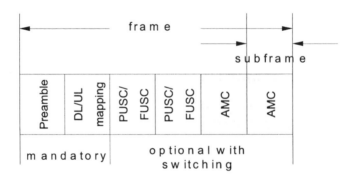

Figure 6.12: Zone switching for multiple traffic channel configurations

Example 36 *The IEEE 802.16e OFDMA mode supports mixed traffic channel configurations with different zones in a frame. As depicted in Figure 6.12, each time frame is partitioned into several subframes, with the first two being the mandatory preamble and the DL- and UL-mapping subframes. The following subframes can be dynamically configured, allowing zone switching between diversity subchannels (e.g., PUSC, FUSC) and coherent subchannels (e.g., AMC). The ratio between diversity subframes and coherent subframes is determined by the application, and primarily the distribution of users' mobility.*

6.4 Summary

In this chapter, we discuss some practical OFDMA system design issues. In particular, we investigate the problem of OFDMA traffic channel configuration, which involves both the PHY and MAC layers. The concept of cross-layer optimization allows MAC and PHY layers to interact based on users' channel pattern and service models. The optimal traffic channel configurations for fixed and mobile applications are derived with reference to the IEEE 802.16e standard. Also described is the scalable OFDMA mode in IEEE 802.16e, where the FFT size scales with the bandwidth size. This scalability offers the advantages of easy implementation and high flexibility in handling user channel/traffic dynamics.

Bibliography

[1] IEEE P802.16e, "Draft Amendment to IEEE Standard for Local and Metropolitan Area Networks Part 16: Air Interface for Fixed and Mobile Broadband Wireless Access Systems, Amendment for Physical and Medium Access Control Layers for Combined Fixed and Mobile Operation in Licensed Bands".

[2] H. Balakrishnan, V. Padmanabhan, S. Seshan and R. H. Katz, "A comparison of mechanisms for improving TCP performance over wireless links," *IEEE/ACM Trans. Networking*, December, 1997.

[3] S. Kunniyur and R. Srikant, "End-to-end congestion control: utility functions, random losses and ECN marks," *IEEE INFOCOM*, vol. 3, March 2000.

[4] G. L. Stuber, *Principles of Mobile Communication*, Boston: Kluwer, 2000.

[5] R. A. Horn and C. A. Johnson, "Matrix Analysis," *Cambridge University Press, Cambridge, England*, 1985.

[6] "Digital video broadcasting (DVB): Framing, channel coding and modulation for digital terrestrial television, Annex A," *ETSI EN300 744 V1.4.1*, 2001.

[7] COST207 TD (86)51-REV 3 (WG1), "Proposal on channel transfer functions to be used in GSM tests late 1986," Sept. 1986.

[8] I. Emre Telatar, "Capacity of Multi-antenna Gaussian Channels," *European Trans. Telecommun.*, vol. 10, pp. 585-595, Nov. 1999.

[9] L. H. Ozarow, S. Shamai and A. D. Wyner, "Information theoretic considerations for cellular mobile radio," *IEEE trans. Information Theory*, vol. 43, May 1994, pp. 359-378.

[10] H. Bolckei, D. Gesbert, and A.J. Paulraj. "On the capacity of OFDM-based spatial multiplexing systems," *IEEE Trans. Commun.*, 50(2):225–234, Feb. 2002.

[11] N. Jindal and A. Goldsmith, "Capacity and optimal power allocation for fading broadcast channels with minimum rates," *IEEE Transaction on Information Theory*, vol. 49, no. 11, Nov. 2003.

[12] B. Varadarajan and J. R. Barry, "The outage capacity of linear space-time codes" submitted, *IEEE Trans. Wireless Commun.*, September 2003.

[13] H. Zhang and T. Guess, "Asymptotical analysis of the outage capacity of rate-tailored BLAST," *IEEE Global Commun. Conf.* (GLOBECOM 2003), San Francisco, CA, Dec. 1-5, 2003.

[14] H. Yaghoobi, "Scalable OFDMA physical layer in IEEE 802.16 Wireless-MAN," *Intel Technology Journal*, vol 8, issue 3, 2004.

[15] T. S. Rappaport, *Wireless Communications: Principles and Practice*, Prentice-Hall PTR, c1996.

[16] ITU-R M.1225, "Guidelines for evaluation of radio transmission technologies for IMT-2000, 1997.

[17] IEEE C802.16d-04_07, "Applying scalability for OFDMA PHY layer."

Chapter 7

Multi-Cell Frequency Planning

7.1 Introduction

The techniques described in previous chapters focus on a single-cell scenario with the assumption that the base station transmits/receives on one set of frequency bands. From a system level perspective, the same set of frequency bands can be reused by multiple base stations, as long as they are physically separated far enough to tolerate the mutual interference. The base stations that use the same frequency band give rise to *co-channel interference*. An important design factor for a cellular system is to coordinate the co-channel interference and make intelligent radio resource allocation over all base stations. The allocation of frequencies to the cellular system is termed *frequency reuse* or *frequency planning* which has a profound impact on system performance [1].

Generally speaking, existing frequency planning schemes can be classified into three categories: static/fixed frequency channel allocation (FCA), adaptive/dynamic channel allocation (DCA) and hybrid channel allocation (HCA). The hybrid channel allocation can be regarded as the combination of FCA and DCA, where some of the channels are fixed for each cell and others are dynamically assigned to cells/users [2] [3].

This chapter first provides an overview of the frequency planning schemes for cellular networks. As will be explained shortly, challenges remain when applying the existing schemes to OFDM systems. A frequency planning scheme based on dynamic channel allocation is then presented for multi-cell OFDMA networks. The scheme takes advantage of mutual interference and channel/traffic diversity. The scheme is associated with a low overhead protocol and a set of low complexity algorithms. The protocol and algorithms are evaluated under different sector configurations and various traffic models. The results reveal some important insights on the trade-off between sector interference suppression and algorithmic interference avoidance.

7.1.1 Fixed channel allocation

In FCA, the total number of channels in the system are divided into disjoint groups and are assigned to a set of cells, with each cell occupying a group of channels. The minimum set of neighboring cells that are assigned the entire set of channels is termed a *cluster* and the number of cells in each cluster is the *cluster size*. As a result, the number of channels each cell occupies is calculated as the total number of channels/cluster size. For a given cell radius, the cluster size determines the *reuse distance,* i.e., the minimum distance between two base stations that can transmit simultaneously without devastating the other's transmission. The relation between the cluster size, reuse distance and cell radius for hexagonal cells can be approximated as [1]

$$rd = \frac{\text{reuse distance}}{\text{cell radius}} = \sqrt{3 \times \text{cluster size}}. \tag{7.1}$$

where rd in (7.1) is called *reuse distance ratio.*

In a given area and a fixed cell size, a smaller cluster size/smaller reuse distance leads to higher capacity (here the capacity is defined as the number of homogeneous users supported simultaneously). Because each cluster supports the same number of channels/users, more clusters lead to more users supported simultaneously. Therefore, in order to cover the same area, more clusters need to be packed into the system if cluster size is smaller, leading to higher capacity. On the other hand, larger cluster size/larger reuse distance lowers the co-channel interference and improves signal to interference ratio (SIR). In practice, the cluster size/reuse distance is determined in the system setup and installation stage and may be adjusted in the system maintenance process.

Many factors have impact on the reuse distance planning. For example, a lower signal to interference requirement (SIR_{req}) for the received signal results in a smaller cluster size and reuse distance. The following example shows how SIR_{req} influences the reuse distance.

Example 37 *Two base stations/cells are in consideration as illustrated in Figure 7.1,. Cell 1 is serving user 1 and cell 2 is serving user 2. If the two BSs transmit simultaneously on the same channel while satisfying the SIR_{req} at the user side, then how far away should the base stations be separated?*

Assume that the received signal power is approximated as

$$p_r = p_0 \frac{p_t}{d^\alpha},$$

where p_t is the transmitter power, d is the distance from the transmitter to the receiver, p_0 is a constant depending on a reference distance and α is the path loss factor. Further assume that the transmission power of BS_1 and BS_2 are the same. Denote the distance from BS_1 to user 1 and from BS_2 to user 1 as d_{11} and d_{21}, respectively, then the received signal to interference ratio (SIR) of user 1 is calculated as

$$SIR = \frac{p_0 p_t / d_{11}^\alpha}{p_0 p_t / d_{21}^\alpha} = \left(\frac{d_{21}}{d_{11}}\right)^\alpha. \tag{7.2}$$

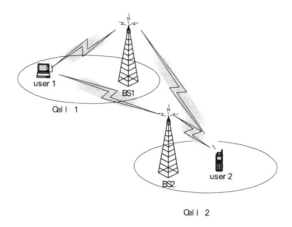

Figure 7.1: Frequency reuse in a two-cell scenario.

In (7.2), it is assumed that the interference dominates the thermal noise at the receiver side. Therefore, there is no noise term in the denominator. In practical wireless systems, an SIR requirement, SIR_{req}, is put on the received signal in order to have a meaningful service. This leads to

$$10 \log_{10} \left(\frac{d_{21}}{d_{11}} \right)^{\alpha} \geq SIR_{req}$$

We then conclude that

$$d_{21} \geq 10^{\frac{SIR_{req}}{10\alpha}} d_{11}. \tag{7.3}$$

(7.3) indicates that the reuse distance depends on the cell radius, the path loss factor and the SIR requirement. A higher SIR_{req} requires a large reuse distance between the base stations while a lower SIR_{req} leads to a smaller reuse distance. For example, when $\alpha = 2$, 10dB SIR requirement leads to a reuse distance of $d_{21} = 3.16d_{11}$, and a 15dB SIR requirement results in $d_{21} = 5.62d_{11}$.

The above example indicates that if the receiver can function in a lower SIR range, then the reuse distance can be decreased, leading to smaller cluster size and higher capacity. Toward this end, tremendous efforts have been taken on interference cancellation/suppression schemes at the physical layer.

Example 38 *One of the interference suppression techniques deployed in GSM networks is single antenna interference cancellation (SAIC) [4]. Handsets with SAIC feature can tolerate higher co-channel interference than regular handsets while maintaining the same level of receiver performance. With SAIC handsets deployed in GSM network, the base stations can transmit with lower power, leading to less interference to other co-channel users. Approximately, the SAIC feature provides 40% capacity increase to the traditional GSM networks [5].*

A good FCA plan also needs to take the traffic loads among cells into account. Unevenly loaded traffic results in unbalanced performance over the cells, which leads to degraded overall system performance. The optimal FCA solution may be an irregular one with unevenly distributed channels over cells. Some variations of FCA accommodate this imbalance with "borrowing" schemes – a cell running out of channels may borrow channels from neighboring cells if it does not violate other cells' transmission.

Generally speaking, FCA puts most effort in the system setup and planning stage, while saving the ease in the system operation process. Therefore, it needs an accurate prediction on the propagation models, traffic loads, user model etc. However, in practice, users' propagation and traffic environments vary in very unpredictable ways. Furthermore, modern wireless networks present a self organized, irregular topology and diverse applications, making these variables even harder to predict. For instance, the SIR requirements may not be a constant in the system, e.g., the requirements for voice applications and multimedia applications are different. These posed challenges further complicate the FCA planning and setup. As a result, certain margins have to be considered in the planning stage. As a result, FCA usually conservatively targets at worst case scenarios, leading to performance loss. On the contrary, DCA handles these dynamic variations more flexibly, leading to more efficient spectrum utilization, especially with low to moderate traffic loads.

7.1.2 Dynamic channel allocation

Unlike FCA, the channel distribution in DCA adapts over time. Instead of predicting and averaging, DCA takes advantage of multiuser channel and traffic diversity to adjust the channel allocation over time. Although DCA requires higher computation complexity and signaling overhead during operation, its ability to utilize real-time system information leads to higher spectrum efficiency, especially with low to moderate traffic loads [6] [7].

Basically, DCA can be divided into two categories: centralized allocation and decentralized/distributed allocation.

- Centralized DCA

  In the centralized DCA scheme [8]-[10], the channel allocation decision is made by a physical/logical central controller, e.g., mobile switch controller (MSC). Although users and BSs are not the decision makers, they do participate in the protocol implementation. The users and BSs are responsible for gathering traffic/channel information and feed back the information to the controller. Since the controller has the global information over the entire network, it yields excellent performance with the cost of intensive signaling overhead and computational complexity.

- Decentralized/distributed DCA

  In the decentralized/distributed DCA scheme [11] [12], the channel allocation decision is made by users or BSs, independently or cooperatively.

The signaling overhead is greatly reduced, and the computation complexity is lowered by distributing the computation load onto different decision makers.

Although DCA has been studied widely in the past decades, there are a few open issues in applying the existing DCA schemes to broadband wireless systems, e.g., OFDMA. The challenges arise from three aspects. First of all, traditional DCA assumes a predetermined SINR threshold for the received signal, which is more suitable for homogeneous services, e.g., voice. Modern data networks employ adaptive coding and modulation, which makes channel assignment decisions non-binary from the SINR standpoint. The transmitters/receivers map different modulation and coding schemes to different SINR values [13]; thus, different throughputs (or achievable rates) are obtained at different SINR levels. Secondly, unlike traditional DCA designed for the simple flat fading environments, modern broadband wireless networks have more variables to account for – users' channels are frequency selective, and their data rate requirements are also different. The third challenge is the intensity of measurement reports, signaling overhead and computation complexity associated with broadband DCA. Since traditional DCA deals with flat fading channels, the overhead due to measurement and signaling is only associated with one frequency band. However broadband networks, e.g., OFDMA, need to exchange information on all of the subcarriers, generating hundreds, even thousands times the overhead of traditional DCA. In addition, the computation complexity will be much higher than the traditional DCA as more channel/traffic variables arise. As a result, fully centralized schemes are often too heavy for implementation as all the interference information on all channels has to be gathered and calculated at a central controller. On the other hand, fully distributed schemes have difficulties dealing with uneven loaded traffic, and hinge upon instantaneous traffic channel establishment [14]. These problems severely complicate the DCA problem for OFDMA systems.

7.2 OFDMA DCA

In this section, we investigate an OFDMA DCA scheme where resource allocation is realized at both a central controller, termed radio network controller (RNC), and base stations (BSs). Our focus is on OFDMA downlink with no intra-cell interference, although the results presented can be extended to uplink applications with minor modifications. The new protocol attempts to capture three types of multiuser diversities, namely, mutual interference diversity, traffic diversity, and selective fading channel diversity. As a consequence, the protocol utilizes three types of key information in a multicell environment: traffic, mutual interference and channel information. Three types of information are exploited by different decision makers– the RNC or BSs, and are utilized based on different time scales – super-frame level or frame level, depending on how fast these diversities can be grasped by the RNC and BSs.

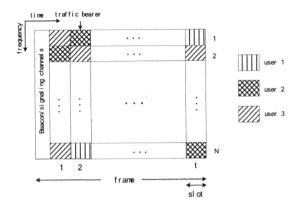

Figure 7.2: OFDMA system in frequecy-time axis. In time domain, one frame is devided into t slots

Specifically, the RNC controls a set of BSs and makes radio resource allocation decisions on a super-frame level to maximize the system throughput and captures mutual interference diversity gain (or interference avoidance gain). This is feasible since mutual interference changes relatively slowly and the interference-floor is within full control of the RNC. The bursty traffic diversity, along with the fading channel diversity is exploited at the BS within shorter frames based on users' changing channel conditions and buffer occupancies. The protocol also accounts for unevenly loaded traffic among cells by satisfying a cell-level QoS constraint at the RNC. The cell-level QoS tries to guarantee that the number of channels to be loaded in different cells is proportional to a pre-set ratio. A set of computationally efficient allocation algorithms for the RNC and BSs are derived along with the protocol. The algorithms perform the functions of both interference avoidance and traffic adaptation with linear-complexity with respect to the number of users and channels. As a result, the channel is always assigned to the user with the highest utility value, which is a function of interference, channel and traffic conditions. The protocol and algorithms are evaluated under different cell configurations with both real-time and non-real-time traffic. The results reveal some important insights on how to trade-off between sector interference suppression and dynamic interference avoidance. In particular, we show that using the proposed DCA scheme, the highest overall system spectrum efficiency can be achieved under four-sector cells with two alternating frequency bands.

7.2.1 Protocol design

Figure 7.2 depicts a typical OFDMA/TDM system, where the radio resource is partitioned in both frequency and time domains. In particular, the frequency resource is divided into L *traffic channels* (each traffic channel is a group of

OFDM subcarriers as described in Chapter 6) whereas the time resource is divided into time slots. The smallest resource unit through which data can be transported is termed *traffic bearer*. Depending on the application, one or a collection of traffic bearers can be allocated to a user at a time. A super-frame is constructed by a number of frames, and a frame is constructed by a number of slots. We invoke the following assumptions for the remainder of this chapter:

• Assumption 1: Each traffic bearer can only be assigned to one user within a given cell, i.e., there is no intra-cell interference.

• Assumption 2: Neighboring cells may reuse the same traffic bearer depending on the mutual interference information.

• Assumption 3: The transmission power on each traffic bearer is fixed, whereas the transmission rate is variable (using adaptive coding and modulation).

• Assumption 4: Only the dominant co-channel signal from neighboring cells is regarded as interference. The rest is treated as background noise.

• Assumption 5: Perfect channel state information at both transmitter and receiver.

Assumption 1 is a natural choice based on Theorem 2 in Chapter 5, which proves that the highest capacity of OFDM system with independent decoding is achieved by assigning each channel only to one user in each cell, i.e., OFDMA. It also shows that water-filling power allocation only brings marginal performance improvement over fixed power allocation with ACM if traffic channels are configured with the maximum multiuser diversity. The use of ACM has a similar effect of power water-filling – good channel quality leads to a higher modulation order and, hence, more transmission power. Similar results have also been reported in [15], which shows that the channel capacity with channel state information at both the transmitter and receiver (i.e., water-filling power allocation) is just marginally larger than that with channel state information only at the receiver (i.e., equal power allocation). Assumption 5 indicates a perfect feedback channel between the transmitter and the receiver, i.e., the feedback channel is error free and has no delay.

We consider an OFDMA system where radio resources are allocated to users based on their channel measurements and traffic requirements. Basically, the RNC coordinates a cluster of BSs. Each BS communicates with a set of users and gathers users' channel state information (CSI) as well as traffic status (e.g., arrival rates, buffer occupancies). Specifically, CSI is defined as a pair of achievable rates which characterizes inter-cell interference and fading channels; please see Figure 7.3 for illustrations. CSI is calculated by users based on the broadcast beacon signals from BSs, and is periodically feed back to BSs. The beacon signals from BSs are used by each user to determine the dominant interfering BS, and the achievable rates with and without the dominant interferer. For

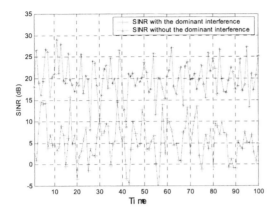

Figure 7.3: SINR variation on a specific traffic channel of a specific user. According to ACM, these two levels of SINR correspond to two transmission rates. As a result, CSI is defined by this pair of rates: CSI=[achievable rate with the dominant interference, achievable rate w/o the dominant interference]

example, on a particular traffic channel, the $SINR$ received by the user who is communicating with BS_i can expressed as

$$SINR = \frac{p_i h_i}{\sum\limits_{j \neq i}^{M} h_j p_j + N_0},\qquad(7.4)$$

where M is the number of BSs in the system and $p_i h_i$ represents the signal strength received by the user from BS_i with p_i and h_i representing the transmitting power of BS_i and channel gain from BS_i to the user, respectively. Using beacon signals, the user can determine the strongest interfering base station, i.e.,

$$k = \underset{j \neq i}{argmax} p_j h_j.$$

The user then calculates its SINR without this dominant interference as

$$SINR = \frac{p_i h_i}{\sum_{j \neq i, j \neq k} p_j h_j + N_0}.\qquad(7.5)$$

The users then map the SINR in (7.4) and (7.5) to the achievable rate with and without the dominant interference. This pair of rates defines the CSI. The actual mapping between the SINR values and the achievable rates is a function of fading profiles and the available ACM techniques; please refer to Chapter 5 for examples of the rate-SINR functions. By defining such a CSI structure, the user's feedback information to BS is reduced to a pair of rates rather than transmitting the signal strength and interference received from M BSs.

Example 39 *In Figure 7.3, the mean SINR with dominant interference is 4dB. If, according to ACM, it corresponds to QPSK modulation with convolutional code rate 1/2, then we obtain an achievable rate of 1 bit/symbol; The mean SINR without dominant interference is 20dB. If, according to ACM, it corresponds to 64QAM with code rate 7/8, then we obtain an achievable rate of 5.25 bits/symbol. In this case, CSI=(1, 5.25)*traffic channel bandwidth. Here the rate is measured by bits/s/traffic channel.*

Such a CSI definition leads to signaling overhead reduction between users and BSs, which will be numerically verified in the ensuing sections. Note that the interference floor can be turned on and off by the RNC, who decides whether the BS is assigned the channel and, thus, becomes the dominant interferer to some user in another cell.

The semi-distributed scheme described here reduces the signaling overhead and computational load by splitting the decisions between the RNC and BSs. Mechanically, the RNC updates all users' CSI from all BSs every super-frame. Decisions made by the RNC include the specific set of traffic channels assigned to each BS for that super-frame and the recommended user assignment for the traffic channel set. Locally, the BS makes the actual pairing between the traffic bearers and the users. In a specific frame, if the recommended user by the RNC has traffic to send, the BS will adopt the RNC's recommendation; otherwise, the BS makes its own decision based on users' traffic conditions (buffer occupancies) and channel fading levels. The decision algorithm of the RNC performs interference avoidance, and the decision algorithm of BSs performs channel/traffic adaptation. Functionally, the RNC is dedicated to coordinate the mutual interference between cells, reducing the information update rate between the RNC and BSs to a super-frame level. BSs exploit the multiuser traffic/channel diversities, making real time decisions on channel assignment at user packet level (frame level). As a result, both the mutual interference diversity and the fading channel/bursty traffic diversity can be efficiently exploited.

7.2.2 Problem formulation for the RNC

The goal of the RNC is to coordinate the mutual interference among BSs. To formulate the channel allocation problem for the RNC, consider an OFDMA system with L traffic channels and a network of M BSs (cells). Denote the number of users in the mth BS as K_m. Thus the entire network has a total of $K_t = \sum_{m=1}^{M} K_m$ users. Let $\Re_m$ denote the user index set for BS$_m$, e.g., $\Re_1 = \{1, ..., K_1\}$, $\Re_2 = \{K_1 + 1, ..., K_1 + K_2\}$. Let us further define the following notations for the rest of the chapter.

- Rate matrices $\mathbf{A}_{K_t \times L} = [a_{kl}]$ and $\mathbf{B}_{K_t \times L} = [b_{kl}]$: a_{kl} and b_{kl} represent user k's achievable rates (bits/s/traffic channel) on channel l with and without the dominant interference, respectively. In other words $[a_{kl}, b_{kl}]$ defines user k's CSI on channel l. In practice, $b_{kl} \geq a_{kl}$.

- Interference index matrix $\boldsymbol{J}_{K_t \times L} = [J_{kl}]$: J_{kl} represents the index of user k's dominant interfering BS on channel l. $J_{kl} \in \{1, 2, ..., M\}$.

- Assignment matrices $\boldsymbol{X}_{M \times L} = [x_{ml}]$ and $\boldsymbol{Y}_{K_t \times L} = [y_{kl}]$: $y_{kl} = 1$ indicates that channel l is assigned to user k and 0 otherwise; $x_{ml} = 1$ indicates that channel l is assigned to BS_m and 0 otherwise.

Define $\delta_{kl} = b_{kl} - a_{kl}$, then $y_{kl} \cdot (b_{kl} - x_{J_{kl}l}\delta_{kl})$ represents the kth user's actual transmission rate on channel l. Thus the total throughput on channel l is:

$$TH_l(\boldsymbol{x}_l, \boldsymbol{y}_l) = \sum_{k=1}^{K_t} y_{kl} \cdot \left(b_{kl} - x_{J_{kl}l}\delta_{kl}\right),$$

where $\boldsymbol{x}_l$ and $\boldsymbol{y}_l$ are the lth column vectors of $\boldsymbol{X}$ and $\boldsymbol{Y}$, respectively. The total throughput of the system (bits/s) is then given by:

$$\begin{aligned}
TH(\boldsymbol{X}, \boldsymbol{Y}) &= \sum_{l=1}^{L} TH_l(\boldsymbol{x}_l, \boldsymbol{y}_l) \\
&= \sum_{l=1}^{L} \sum_{k=1}^{K_t} y_{kl} \cdot \left(b_{kl} - x_{J_{kl}l}\delta_{kl}\right).
\end{aligned}$$

Since each channel is assigned to only one user at any time within BSs, we obtain the following relationship:

$$x_{ml} = \sum_{k \in \Re_m} y_{kl}.$$

As a result, $\boldsymbol{X}$ is uniquely determined by $\boldsymbol{Y}$, and the total throughput can be re-expressed as a function of only $\boldsymbol{Y}$:

$$\begin{aligned}
TH(\boldsymbol{Y}) &= \sum_{l=1}^{L} TH_l(\boldsymbol{y}_l) \\
&= \sum_{l=1}^{L} \sum_{k=1}^{K_t} y_{kl} \cdot \left(b_{kl} - \sum_{i \in \Re_{J_{kl}}} y_{il}\delta_{kl}\right).
\end{aligned}$$

The throughput on a specific channel l is now expressed as:

$$TH_l(\boldsymbol{y}_l) = \sum_{k=1}^{K_t} y_{kl} \cdot \left(b_{kl} - \sum_{i \in \Re_{J_{kl}}} y_{in}\delta_{kl}\right). \tag{7.6}$$

Each cell may serve different types of users. Therefore their throughput requirements and demands vary. To account for this matter, a cell-level QoS constraint is posed to require that the number of channels assigned to cells is proportional to a pre-set ratio. The RNC seeks to find the assignment matrix

$\mathbf{Y}$ so that the total throughput is maximized while satisfying the cell-level QoS requirement. Mathematically, the RNC problem is formulated as:

$$\underset{\mathbf{Y}}{Max}\ TH(\mathbf{Y}) = \sum_{l=1}^{L}\sum_{k=1}^{K_t} y_{kl} \cdot \left(b_{kl} - \sum_{i \in \Re_{J_{kl}}} y_{il}\delta_{kl} \right) \tag{7.7}$$

$$subject\ to\ 1)\ \sum_{k \in \Re_m} y_{kl} \in \{0,1\},\ m = 1, 2, ..., M; l = 1, 2, ...L$$

$$2)\ |\Phi_1| : |\Phi_L| = \Theta_1 : ... : \Theta_L$$

$$3)\ y_{kl} \in \{0,1\},\ k = 1, 2, ..., K_t; l = 1, 2, ...L.$$

The first constraint in (7.7) is to guarantee that at most one user is using the channel within a BS on every channel. The second constraint is to satisfy the cell-level QoS requirement. Φ_l denotes the channel set assigned to BS$_l$ and $|\Phi_l|$ represents the cardinality of the set, i.e., the number of channels in Φ_l. Θ_l is the pre-set ratio for the number of channels to be assigned to cells. For example, for two cells with different number of users that have similar traffic models, the ratio can be set as the ratio of the number of users in the cells. The assignment for cells, $\mathbf{X}$ is then uniquely determined by the resulting $\mathbf{Y}$. In the protocol, $\mathbf{X}$ will be sent to BSs together with $\mathbf{Y}$ to indicate the channel assignment decisions for BSs and the recommended user on each assigned channel.

7.2.3 Problem formulation for the BSs

After the set of channels are allocated to a BS, the objective of the BS is to capture the traffic and fading channel diversity within the current super-frame. If the recommended user has traffic to send, then the BS will obey the RNC's decision. However, for bursty traffic, the recommended user by the RNC may not have traffic to send in each frame. In this case, the BS will make its own decision to maximize the system throughput based on the users' fading channel and traffic conditions. In this case, the traffic condition is represented by the buffer occupancies, i.e., the number of bits waiting in the user's buffer to be transmitted. The buffer occupancy characterizes the user's traffic intensity and requirement.

Taking BS$_1$ for an example, assume L_1 channels are assigned to the BS, and it has K_1 users. Since the interference level for its users is pre-determined during each super-frame (i.e., $\mathbf{X}$ is fixed for the current super-frame), the achievable rate of each user is uniquely given by $\mathbf{U}_{K_1 \times L_1} = [u_{kl}]$ where $u_{kl} = (b_{kl} - x_{J_{kl}l}\delta_{kl})$, indicating user k's achievable rate on channel l.

Let $\mathbf{Z}_{K_1 \times L_1} = [z_{kl}]$ be the channel assignment matrix for BS$_1$, with $z_{kl} = 1$ indicating that channel l is assigned to user k and 0 otherwise. Then the number of bits user k can transmit in one slot is given by

$$r_k = t_s \sum_{l=1}^{L_1} u_{kl}z_{kl},$$

where t_s is the duration of one slot and it is regarded as a constant in the system. Again, channel l is assigned to only one user at any time, thus the following constraint must be satisfied:

$$\sum_{k=1}^{K_1} z_{kl} \in \{0, 1\}, l = 1, 2, ..., L_1.$$

In each slot, the BS has the knowledge of each user's buffer occupancy which is expressed as

$$c = (c_1, c_2, ..., c_{K_1}).$$

Based on traffic condition c and channel condition U, the BS seeks to find a channel assignment matrix Z for each slot in the current frame so as to maximize the total throughput. Note that the number of bits user k can send in one slot is $\min\{c_k, r_k\}$. Also note that r_k is a function of Z, so the total throughput in each slot TH' becomes a function of Z:

$$TH'(Z) = \sum_{k=1}^{K_1} \min\{c_k, r_k\}.$$

As a result, the BS problem is formulated as:

$$\underset{Z}{Max}\ TH'(Z) = \sum_{k=1}^{K_1} \min\{c_k, r_k\}$$

$$subject\ to\ 1) \sum_{k=1}^{K_1} z_{kl} \in \{0, 1\}, l = 1, 2, ..., L_1, \tag{7.8}$$

$$2) z_{kl} \in \{0, 1\}, l = 1, 2, ..., L_1, k = 1, 2, ...K_1.$$

The problems formulated for the RNC and BSs are both nonlinear integer optimization problems and are essentially very difficult to solve [16]. Cutting plane and branch-and-bound algorithms have been suggested to deal with certain classes of nonlinear integer programming problems. However, there is no guarantee that these algorithms yield good performance over large scale problems (e.g., problems with a few dozens of constraints and variables like the RNC and BS problems) [17]. For these reasons, we derive two suboptimal algorithms for the RNC and BS problems, respectively. The algorithms fall into the category of the "local search" method, which has been shown to be very effective in a variety of nonlinear integer optimization problems.

7.2.4 Fast algorithm for the RNC

The RNC algorithm uses a greedy approach and attempts to assign a traffic channel to the BS that has the highest positive *throughput marginal utility* (TMU) value (denoted as Ω_k in the algorithm). TMU is defined as the system throughput improvement by assigning the current channel to the user within the BS being evaluated. The channel assignment is progressively performed to provide the most improvement to the system throughput. If none of the users in the BS has a positive TMU value, then the channel is not assigned to this particular BS. The algorithm performs channel allocation as summarized in Table 7.1.

The RNC algorithm

Inputs: A, B, J

Outputs: X, Y

Initialization: X=[ ]; Y=[ ];

For $l = 1 : L$ do   % start of channel loop

   $x_{M \times 1} = (0, 0, ..., 0)^T$;   % initialize the lth column vector of X

   $y_{K_t \times 1} = (0, 0, ..., 0)^T$;   % initialize the lth column vector of Y

   $\pi = sort(\frac{size(\Phi_m)}{\Theta_m})$;

    % sort BSs based on the pre-set ratio and the assigned channels.

   For $j = 1 : M$ do   % start of BS loop

     $m = \pi(j)$;   %start from the BS that is most under-assigned

     For $k \in K_m$ do   % start of user loop

      $\Omega_k = TH_l(y + e_k^{K_t}) - TH_l(y)$;

      % calculate the TMU value of each user in BS$_m$ using (7.6)

     End For   % end of user loop

     $k^* \leftarrow \arg\max_k \Omega_k$;   % find the user with the highest TMU

     If $\Omega_{k^*} > 0$

     % if assigning channel l to BS$_m$ can improve the total throughput

      $x = x + e_m^M$;   % assign the channel to BS$_m$

      $y = y + e_{k^*}^{K_t}$;   % indicated the recommended user

      $\Phi_m = \Phi_m \cup \{l\}$ % add channel l to channel set of BS$_m$

     End If

   End For   % end of BS loop

   $X = [X \ x]$;   % assign x to the lth column of X

   $Y = [Y \ y]$;   % assign y to the lth column of Y

End For   % end of channel loop

Table 7.1: RNC algorithm

In the RNC algorithm, $e_k^{K_t}$ is a K_t-length vector with all elements being 0 except the kth element being 1. Φ_m is the set of channels assigned to BS$_m$. π is the vector containing the order with which BSs are evaluated. π is determined based on the pre-set throughput ratios $\{\Theta_m\}$ for BSs and the number of channels already assigned to BSs.

Essentially, the algorithm assigns the channels one by one. For each channel it calculates Ω_k, the TMU value, for each user. Ω_k represents user k's contribution to system throughput improvement due to the higher utilization of the channel minus the effect of the throughput loss due to the increased mutual interference introduced by this user/BS. The algorithm is carried out in three loops. For each channel loop, the evaluation is executed in BS loop and user loop. Clearly, the order with which BSs are evaluated has a profound impact on the final output. The first BS usually gets the "clear" channel whereas the rest can only use it if the TMU value is positive. It is possible for Ω_k to be negative for all the remaining users. This means that due to the interference from previous assignments, no more BS can increase the total throughput by using this particular channel. In this case, the channel will not be assigned to any additional BS. To address the evaluating priorities of BSs, the order of BS loop is adjusted after each channel assignment. More specifically, the BS that is most under-assigned ($\arg\min\limits_{m} \frac{size(\Phi_m)}{\Theta_m}$) will be picked first and the BS that is most over-assigned ($\arg\max\limits_{m} \frac{size(\Phi_m)}{\Theta_m}$) will be examined last in BS loop.

7.2.5   Fast algorithm for BSs

Once each BS receives its channel assignment from the RNC within a particular super-frame, it will then make instantaneous decisions on pairing the traffic bearers and users. Note that the recommended user by the RNC may not always have traffic to send in each frame. In this case, the BS seeks to solve (7.8) for each slot in current frame. Note that once the RNC has made decisions on which channel is used by which BS, the mutual interference from BSs to users is pre-determined for that super-frame. Therefore, re-allocating the channel to a user who has traffic to send, instead of maintaining the RNC's recommended user on that channel who has an empty queue, will always improve the throughput for the BS. In this case, the RNC algorithm does not remain optimal in this frame, since the BS does not pair channels and users according to the RNC's decision. However, to avoid back-and-forth information and decision exchange between BSs and the RNC in each slot, we still allow BSs to make their own decisions.

The algorithm introduced in [18] is modified to solve (7.8). Note that if channel l is assigned to user k, the number of bits that can be transmitted using channel l in one slot is

$$F_k = \min(c_k,\ t_s \cdot u_{kl}).$$

F_k is defined as the *data traffic utility* (DTU) value, and it represents how much throughput user k can obtain using the channel in one slot. F_k captures the traffic condition (c_k) as well as the fading channel characteristic (u_{kl}). The BS algorithm attempts to find the user that has the highest DTU value for each traffic bearer and makes the most use of each traffic channel. The algorithm is carried out for each slot in the frame and it is described in Table 7.2, taking BS_1 as an example.

As a result, the channel assigned to the user not only has good channel condition but is also guaranteed to be utilized indeed.

BS Algorithm

Inputs: U and c

Output: Z

For $l = 1 : L_1$ do

$\quad F_k = \min(c_k, t_s \cdot u_{kl})$, $k = 1, 2, ..., K_1$; % calculate users' DTU values

$\quad k^* \leftarrow \underset{k}{argmax}\ F_k$ % find the user that can make the most use of channel l

$\quad z_{k^* l} = 1;$ % assign channel l to the user

$\quad c_{k^*} = c_{k^*} - t_s \cdot u_{k^* l}.$ % re-calculate the user's queue length for the next slot

End For

Table 7.2: BS algorithm

Proposition 13 *The RNC algorithm has complexity of $O(L \times M \times K_t)$, and the BS algorithm has complexity of $O(L_1 \times K_1)$.*

Proof. Please see Appendix for the proof.   ■

7.3 Spectrum efficiency under different sector configurations

We study the performance of the protocol and the allocation algorithms by simulating a multicell, frequency-division duplexing (5MHz + 5MHz) OFDMA system. The 5MHz band is then divided into OFDMA traffic channels. The following algorithms are evaluated:

1. *RAND*, which randomly allocates traffic channels to users and each traffic channel is reused in all BSs.

2. *RANDBS*, which randomly allocates traffic channels to users and each traffic channel is reused in all BSs. However when an assigned user does not have data to send, the BS assigns the channel randomly to another user with traffic.

3. *RNC*, which only performs the RNC's interference avoidance algorithm. The BSs obeys the RNC's decision all the time.

4. *RNCBS*, which performs both the RNC and BS algorithms.

The performance gains are quantified under two categories: interference avoidance gain (IA) and traffic diversity (TD) gain. The algorithm implemented by the RNC captures IA gain and partial fading diversity gain since no traffic feature is accounted. The IA and partial fading diversity gain exploited by the RNC can be quantified by the performance comparison between *RNC* and *RAND*. On the other hand, the algorithm implemented by BSs exploits both traffic diversity gain and some fading diversity gain. The TD gain and partial

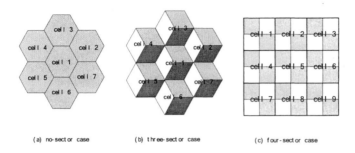

Figure 7.4: Sectorization configuration

fading diversity gain exploited by BSs can be quantified by the performance comparison between *RNC* and *RNCBS*, and between *RAND* and *RANDBS* under bursty traffic configurations.

7.3.1 System configuration and signaling overhead

We consider three types of sector configurations: no-sector (using omnidirectional antennas), three-sector (using 120° antennas) and four-sector (using 90° antennas). Users are uniformly distributed in each sector. Antenna gains are generated according to [19]. A total of 7 cells are simulated for no-sector and three-sector cases, while 9 cells are simulated in four-sector cases as shown in Figure 7.4. Each color represents a 5MHz band and is coordinated by the RNC. Note that in the four-sector configuration, a traffic channel is reused in two sectors within the same cell.

Two types of data services, real-time and non-real-time traffic, are studied in simulations. The real-time service is modeled as a constant rate data stream with 100% activity [19]. The non-real-time service is simulated using the packet train model [20]. Pareto distribution is used to generate individual traffic: $Prob(ON/OFF\ period > t) = (\frac{t_{min}}{t})^\alpha$, where the mean ON/OFF duration is $\frac{\alpha}{\alpha-1}t_{min}$. In all simulations, we set $\alpha = 1.7$ and mean=7.2 s in ON period, and $\alpha = 1.2$ and mean=10.5 s in OFF period.

Path loss is calculated based on the model proposed by Vinko et al [21]:

$$PL = PL_0 + 10(\alpha - \beta \cdot \gamma_b + \theta/\gamma_b)log_{10}(d/d_0),$$

where α, β, θ and γ_b are type 2 model parameters (hilly/light tree density or flat/moderate-to-heavy tree density) and PL_0 is the reference path loss at d_0. In all simulations, $d_0 = 1Km$. Shadow fading with a shadowing variance of 8 dB is assumed:

$$Sh\_new = C(d) * Sh\_old + \sqrt{1 - C^2(d)}Gaussian(0, \sigma),$$
$$C(d) = e^{-\frac{d}{d_r}\ln 2},$$

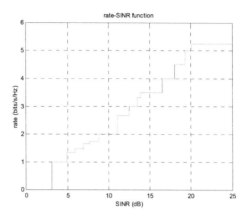

Figure 7.5: Adaptive coding/modulation scheme used in simulations

where Sh_old represents the shadowing value of the last calculation. Sh_new is the current shadowing value. $Gaussian(0, \sigma)$ stands for a Gaussian variable with mean 0 and variance $\sigma^2 = 8^2 = 64$ dB. $C(d)$ determines the correlation of shadowing values between calculations. d_r is a reference distance which is set to 5m and d is the difference of the user's positions between adjacent calculations. For simplicity, users' positions are fixed while d is calculated according to the Doppler spread used in simulations (the default Doppler spread is 10Hz). Fast fading is modeled with the filtered-noise model and frequency selectivity is generated using COST207-TU model [7]. The rate-SINR function of the ACM scheme is shown in Figure 7.5. The achievable rates in CSI are obtained by multiplying the corresponding rate with the traffic channel bandwidth.

The basic OFDMA setup is given in Table 7.3. The signaling overhead is calculated in the following example.

Example 40 *For each traffic channel and each user we use 6 bits to represent its CSI and 3 bits for the identification of the dominant interfering BS. The amount of information sent by all the users is thus $(6+3)$ bits/user/channel×32 channels×100 users=28.8 Kbits in uplink. Since the information is updated every super-frame, the uplink signaling rate is 28.8 Kbits/600ms=48 Kbps. In downlink, we use 7 bits to represent the user index for every traffic bearer. The signaling rate is thus 7 bits/channel/slot×32 channels×7 slots/10ms = 156.8 Kbps. If we assume a modest 1 bit/s/Hz bandwidth efficiency on average, the real traffic throughput for uplink and downlink are (5.12 Mbps-48 Kbps)= 5.07 Mbps and (5.12 Mbps-156.8 Kbps)= 4.96 Mbps, respectively. The overall overhead is quite reasonable as summarized in Table 7.4.*

number of subcarriers	512
subcarrier bandwidth	10 KHz
sys. bandwidth	10K*512=5.12 MHz
number of traffic channels	32
number of active users	100 users/sector
super-frame length	600 ms (default)
frame length	10 ms
slots per frame	7

Table 7.3: OFDMA system parameters

Information Type	Traffic Through-put	Signaling Over-head	Overhead Per-centage
uplink	5.07 Mbps	48 Kbps	0.94%
downlink	4.96 Mbps	156.8 Kbps	3.06%

Table 7.4: Signaling overhead

Interference avoidance gain (IA gain)

We define the spectrum efficiency as

$$E_s = \frac{\sum_{l=1}^{L} \sum_{m=1}^{K_t} y_{kl} \cdot (b_{kl} - x_{J_{kl}l} \delta_{kl})}{\text{total bandwidth}}.$$

E_s is the total achievable rate (summed over all users on all channels) averaged by the system bandwidth. The allocation matrix X and Y are obtained from the allocation algorithms. The re-use percentage is defined as the percentage of cells that can re-use the same channel simultaneously. The following example compares $RAND$ (without IA gain) and RNC (with IA gain).

Example 41 *Table 7.5 summarizes the performance in different sector config-urations. The results are obtained by averaging over 5000 super-frames and all traffic channels and users. Using the RNC algorithm, the improvements due to IA gain (improvement in E_s) are 70%, 27% and 37% for no-sector, three-sector and four-sector configurations, respectively. As expected, the RNC algorithm is most efficient in scenarios with heavy inter-cell interference (no-sector case). The gain decreases in the three-sector case where most interference is suppressed by sector antennas.*

With or without interference avoidance, sectorization improves the overall system spectrum efficiency E_s. The highest E_s is obtained in the four-sector setup since it has two sectors re-using the same frequency simultaneously. Also note that due to the RNC coordination, over 75% of the traffic channels are reused by all cells at all times.

Next, we define the system utilization as follows:

	no-sector (w/o IA / IA)	3-sector (w/o IA / IA)	4-sector (w/o IA / IA)
Re-use Percentage	100% / 85%	100% / 93%	100% / 75%
E_s	2.06 / 3.50 (bits/s/Hz)	3.09 / 3.95 (bits/s/Hz)	3.3 /4.5 (bits/s/Hz)

Table 7.5: Spectrum efficiency in difference sector cases

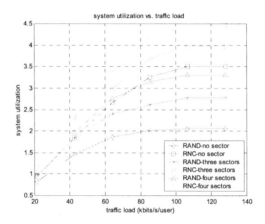

Figure 7.6: System utlization vs. traffic load

$$\text{system utilization} = \frac{\text{total throughput of all users and channels}}{\text{system bandwidth}}.$$

Clearly the system utilization is a function of the traffic load: higher traffic load tends to yield higher throughput. However as traffic load increases, the system utilization converges to the system spectrum efficiency E_s.

Example 42 *Figure 7.6 shows the system utilization vs. traffic load with different sectorizations. Among all configurations, the four-sector setup has the highest system utilization by balancing sectorization gain and the interference avoidance gain.*

7.3.2   Channel loading gains

Traffic diversity gain (TD gain)

To measure the traffic diversity gain and the partial channel fading diversity gain exploited by BSs, we now consider non-real-time services, where traffic

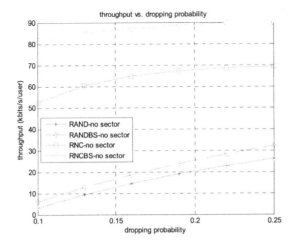

Figure 7.7: Average throughput per user vs. dropping probability using different algorithms in no-sector case

arrives at BSs in a bursty fashion. We define dropping probability as:

$$\text{dropping probability} = \frac{\text{dropped bits due to buffer overflow}}{\text{total number of bits for transmission}}$$

In simulation, we set buffer size to $2K$ bits for each user.

Example 43 *Figure 7.7 gives average throughput per user vs. dropping probability. RNCBS gives the highest throughput under the same dropping probability while RAND being the worst. The better performance of RNCBS is due to the interference avoidance scheme used by the RNC algorithm as well as the traffic diversity gain captured by the BS algorithm. Note that the difference between RNCBS and RNC is much greater than the difference between RANDBS and RAND. This is explained by the fact that the BS in RNCBS performs the BS algorithm and re-assigns channels based on users' traffic and channel conditions, while the BS in RANDBS only re-assigns channels randomly. In other words, the gap between RAND and RANDBS quantifies partial TD gain, while the gap between RNCBS and RNC quantifies TD gain as well as part of the fast fading diversity gain.*

Channel re-assignment impact

One of the key issues in system design is the rate at which resource re-allocation should be performed.

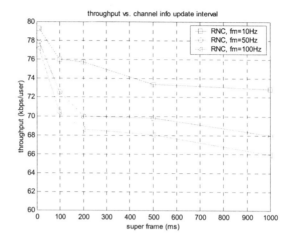

Figure 7.8: Average throughput per user vs. channel re-assignment interval under three Doppler spreads: 10Hz, 50Hz and 100Hz

Example 44 *Figure 7.8 studies the system throughput under different channel variation conditions and different RNC re-allocation frequencies. The superframe length varies from 10 ms to 1s and the channel fading is studied under three Doppler spreads: 10Hz, 50Hz and 100Hz.*

As expected, slower channel variation yields higher throughput, while more frequent re-allocation update brings higher throughput. Since the RNC algorithm captures IA gain and also partial fading diversity gain, slower varying channels (or faster re-allocation) allows the RNC to capture more accurate information and make better decisions. The throughput drop due to longer superframes (or faster varying channels) quantifies the fading diversity gain relative to the total multiuser diversity gain. It is seen that under very fast fading (e.g., when Doppler spread is 100 Hz), the throughput degrades by 15% compared to that under relatively slower fading, e.g., when Doppler spread is 10 Hz. In reality, only a very small portion of the broadband users will be highly mobile. Therefore, it is reasonable to expect an even smaller drop in overall system throughput.

7.4 Summary

In this chapter, we first provide an overview of the frequency planning schemes in wireless networks. Next, we consider the DCA scheme in OFDMA networks and describe a semi-distributed radio resource control scheme, where DCA is realized at both the RNC and BSs. To capture the multiuser diversity in a multicell broadband environment, the RNC coordinates intercell interference over BSs at the super-frame level, where each BS makes its channel assignment decision on the frame level based on the users' channel as well as traffic conditions. As a result, the radio resources are always assigned to the user with the highest utility value. Two suboptimal channel allocation algorithms are presented for the RNC and BSs, respectively. The spectrum efficiency of multicell OFDMA systems under different cell/sector configurations is evaluated.

Appendix: Proof of Proposition 13

Definition 8 *Time complexity: Let A be an algorithm which accepts inputs from a set Ξ, and let $f : \Xi \to R^+$. If there exists a constant $\alpha > 0$ such that A terminates its computation after at most $\alpha f(\theta)$ elementary steps (including arithmetic operations) for each input $\theta \in \Xi$, then we say that A runs in $O(f)$ time. We also say that the running time (or the time complexity) of A is $O(f)$.*

Definition 9 *Polynomial running time: An algorithm with rational inputs is said to run in polynomial time if there is an integer k such that it runs in $\alpha O(n^k)$ time, where $\alpha > 0$ and $\alpha < \infty$, where n is the input size, and all numbers in intermediate computations can be stored with $O(n^k)$ bits. In the case $k = 1$, we have a linear-time algorithm, which is regarded as efficient in general.*

Proposition 14 *The RNC algorithm runs in polynomial time with $f = N \times L \times M_t$, and $k=1$, i.e., it is a linear-time algorithm.*

Proof. Let $\Xi_1 = R_+^{M_t \times N}$ and $\Xi_2 = \{1, 2, ...L\}^{M_t \times N}$; thus, $\mathbf{A}, \mathbf{B} \in \Xi_1$ and $\mathbf{J} \in \Xi_2$. The input for the RNC algorithm is $\Xi = \Xi_1 \times \Xi_1 \times \Xi_2$ and the size of Ξ is $3M_t \times N$. For each input $\theta \in \Xi$, the RNC algorithm has three loops: the channel loop, the BS loop and the user loop. For each channel loop, the algorithm has L iterations. Let each iteration time be $\alpha M_t + \beta$ where α and β are the number of elementary steps needed for calculating Ω_m and the rest of the iteration, respectively. Since $\alpha, \beta < \infty$ (we can make such an assumption because calculations of Ω_m and the rest steps of the iteration have fixed and finite number of elementary steps), the time complexity of the RNC algorithm is $N \times O(L \times (\alpha M_t + \beta)) = O(N \times L \times M_t)$. Therefore the RNC algorithm runs in $O(f)$ time with $f = N \times L \times M_t$.

Note that $O(N \times L \times M_t) = L \times O(3M_t \times N) = L \times O(sizeof (\Xi))$. Further assume that the intermediate steps can be stored with $L \times O(size\ of\ (\Xi))$ bits (we can make such an assumption because the intermediate steps X and Y are of size $L \times N$ and $M_t \times N$. As elements of X and Y are 0 or 1, we can store both of them in less than $3M_t \times N$ bits), we conclude that the RNC algorithm is a linear-time algorithm. Note that, in general, the input size is the the number of bits needed to represent all the entries of the inputs. Here for simplicity, the dimension of the inputs is used as the input size. The proofs can be easily extended to the general case.   ∎

Proposition 15 *The BS algorithm runs in polynomial time with $f = N$ and $k=1$, i.e., it is also a linear-time algorithm.*

Proof. Take BS_1 as an example. Let $\Xi_1 = R_+^{M_1 \times N_1}$ and $\Xi_2 = R_+^{M_1}$; thus, $U \in \Xi_1$, $c \in \Xi_2$ and the input of the BS algorithm is $\Xi = \Xi_1 \times \Xi_2$. For each input $\theta \in \Xi$, BS algorithm has N_1 iterations. Let each iteration time be $\alpha < \infty$ (We can make such assumption because each step of the iteration costs finite and fixed number of elementary steps), so the running time complexity of the BS algorithm is $O(N_1)$. Therefore, the BS algorithm runs in $O(f)$ time with $f = N_1$. Since $O(N_1) < O(M_1 \times N_1 + M_1) = O(sizeof\ (\Xi))$, similar to RNC algorithm, we can assume that the intermediate steps can be stored in $O(sizeof\ (\Xi))$ bits, then the BS algorithm is also a linear-time algorithm. ∎

Bibliography

[1] T. S. Rappaport, *Wireless communications: principles and practice*, Prentice-Hall PTR, 1996.

[2] T. J. Kahwa and N. Georganas, "A hybrid channel assignment scheme in large scale cellular structured mobile communication systems, " *IEEE Trans. Commun.*, vol. COM 26, 1978, pp. 432-438.

[3] D. Cox and D. Reudink, "Increasing channel occupancy in large scale mobile radio systems: dynamic channel reassignment, " *IEEE Trans. commun.*, vol. 21, 1973, pp. 1302-1306.

[4] 3GPP Specification 45.903, available online at URL: Http://www.3gpp.org/ftp/specs/.

[5] 3GPP Document GP-040409, available online at URL: Http://www.3gpp.org/ftp/tsg_geran/.

[6] I. Katzela and M. Naghshineh, "Channel assignment schemes for cellular mobile telecommunication systems: a comprehensive survey," *IEEE Personal Communication*, June 1996. pp. 10-31.

[7] R. Beck and H. Panzer, "Strategies for handover and dynamic channel allocation in micro-cellular mobile radio systems," *IEEE Proc. VTC*, San Francisco, 1989.

[8] D. C. Cox and D. O. Reudink, "Dynamic channel assignment in two dimension large-scale mobile radio systems," *Bell Sys. Tech. Journal*, vol. 51, 1972.

[9] M. Zhang, "Comparison of channel assignment strategies in cellular mobile telephone systems, " *IEEE Trans. Vehicular Tech.*, vol. VT38, 1989, pp. 211-215.

[10] K. Okada and F. Kubota, "On dynamic channel assignment in cellular mobile radio systems," *IEICE Trans. Fundamentals*, vol. 75, 1992, pp. 938-941.

[11] C. L. I and P. H. Chao, "Local packing – distributed dynamic channel allocation at cellular base station, " *IEEE GLOBECOM*, 1993.

[12] F. Furuya and Y. Akaiwa, "Channel segregation, a distributed adaptive channel allocation scheme for mobile communication systems," *Trans. IE-ICE*, vol. E74, June 1991, pp.1531-1537.

[13] A. Czywjk, "Adaptive OFDM for Wideband Radio Channels," *IEEE Global Telecommun. Conf.*, London, vol. 1, pp. 713 -718, Nov. 1996.

[14] J. Chuang and N. Sollenberger, "Beyond 3G: Wideband wireless data access based on OFDM and dynamic packet assignment," *IEEE Commun. Mag.*, vol. 38, July 2000, pp. 78-87.

[15] E. Biglieri, J. Proakis and S. Shamai, "Fading channels: Information-theoreric and communications aspects," *IEEE Trans. Inform. Theory*, vol. 44, Oct. 1998, pp. 2619-2692.

[16] B. Korte, J. Vygen and J. Vygen, *Combinatorial Optimization: Theory and Algorithms*, Second Edition, Springer-Verlag 2002.

[17] C. H. Papadimitriou and K. Steiglitz, *Combinatorial Optimization: Algorithms and Complexity*, Prentice-Hall, Englewood, Cliffs, NJ, 1982.

[18] G. Li and H. Liu, "Dynamic resource allocation with buffer constraints in broadband OFDMA networks," *WCNC'2003*, March, 2003, New Orleans.

[19] ETSI document TR 101.112 V3.2.0, 3G document UMTS 30.03 v3.2.0, available online at URL: Http://www.3gpp.org/ftp/specs/.

[20] W. Willinger, M. S. Taqqu, R. Sherman and D. V. Wilson, "Self-similarity through high-variability: statistical analysis of ethernet LAN traffic at the source level," *IEEE Trans. Networking*, vol. 5, no. 1, Feb. 1997, pp. 71 -86.

[21] V. Erceg, L. J. Greenstein, S. Y. Jjandra, S. R. Parkoff, A. Gupta, B. Kulic, A. A. Julius and R. Bianchi, "An empirically based path loss model for wireless channels in suburban environments," *IEEE JSAC*, vol. 17, no. 17, July, 1999, pp. 1205-1211.

Chapter 8

Appendix

8.1 IEEE 802.11 and WiFi

8.1.1 802.11 overview

In recent years, the demand for rendering multimedia applications over wireless has motivated the development and enhancement of IEEE 802.11 wireless local area network (LAN). Compared to the traditional Ethernet LAN, Wireless LAN has the merits of easy installation, low cost and supporting certain degree of mobility. 802.11 is a part of the 802 standard family for local area networks. This family defines the physical and data link layer specified in the International Organization for Standization (ISO) Open Systems Interconnection (OSI) basic reference model. More specifically, 802.11 defines the medium access control (MAC) layer and physical (PHY) layer. Its relation with other IEEE 802 standards is illustrated in Figure 8.1.

The illustrated standards in Figure 8.1 are described as follows:

- 802– Overview and Architecture. This standard is an overview of the

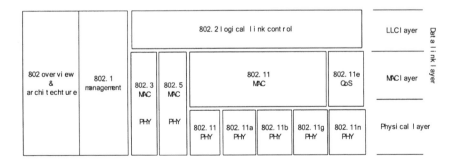

Figure 8.1: IEEE 802 standard family for local area networks

IEEE 802 standard family.

- 802.1B/1E/1F– LAN and MAN Management/ System Load Protocols/ Common Definitions and Procedures. This set of standards define LAN/ MAN management related services, protocols, architectures and procedures.

- 802.2– Logical Link Control. This standard defines the Logic link control sublayer specification.

- 802.3– CSMA/CD Access Method and Physical Layer Specifications. This standard defines the Ethernet MAC layer and PHY layer specifications.

- 802.5– Token Ring Access Method and Physical Layer Specifications. This standard defines the MAC layer and PHY layer for Token Ring Networks.

- 802.11– Wireless LAN MAC and PHY Specifications. This is the first 802.11 standard which defines the CSMA/CA MAC scheme and three PHY schemes: Infrared, frequency hopping spread spectrum and direct sequence spread spectrum.

- 802.11a – Wireless LAN MAC and PHY Specifications: High Speed PHY Layer in the 5 GHz Band. This standard defines the OFDM PHY specification operating at "Unlicensed national information infrastructure" (U-NII) 5 GHz frequency band.

- 802.11b – Wireless LAN MAC and PHY Specifications: High-Speed PHY Layer Extension in the 2.4 GHz Band. This standard is an enhancement of the 802.11 PHY layer in the 2.4 GHz band. It is backward compatible with 802.11.

- 802.11g – Wireless LAN MAC and PHY Specifications: Further Higher Data Rate Extension in the 2.4 GHz Band. This standard is a further enhancement of 802.11 and 802.11b. It is backward compatible with 802.11 and 802.11b.

- 802.11e – This standard has not been finalized yet and is handled by 802.11 Task Group (TG) e. The TG e is working on the enhancement of 802.11 MAC layer to provide better quality of service in 802.11 network. The finalized standard shall be named as 802.11e.

- 802.11n – This standard has not been finalized yet. It is currently handled by 802.11 Task Group n. The TG n is responsible for the physical layer technologies for future wireless LAN supporting minimum 100 Mbps MAC data rate. The finalized standard shall be named as 802.11n.

The last letter in the standard name indicates the Task Group that is responsible for improving certain aspects of the standard. For instance, 802.11 task group A is working on wireless LAN standard in the 5 GHz band. The standard approved by Task Group A is named as 802.11a.

Referring to Figure 8.1, 802.11 standard family (802.11 and its enhancements 802.11a/b/g/n) is a set of MAC and PHY specifications that work with the LLC encapsulation defined in 802.2. In fact, the main difference of these standards is their physical layer technologies; their MAC layer scheme remains unchanged.

The first version of 802.11 was approved in 1997. It adopts three types of PHY options: frequency hopping spread spectrum (FHSS), direct-sequence spread spectrum (DSSS) and Infrared (IR) techniques, though the last one is not widely deployed. By then, the maximum data rate is 2 Mbps. 802.11 was initially designed in the 2.4 GHz unlicensed "industrial, scientific and medical" (ISM) band. This band of frequencies is also crowded with microwave and several other systems. People then began to seek other frequency band options. In 1999, the second version, 802.11a, was approved, and it is operating in "unlicensed national information infrastructure" (U-NII) 5 GHz band. 802.11a adopts OFDM as its physical layer technology. The new technology of 802.11a PHY boosts the data rate from 2 Mbps to 54M bps. 802.11b was also approved in 1999, and it is still operating in the 2.4GHz band. The 802.11b PHY enhances the 802.11 DSSS PHY scheme with HR-DSSS PHY scheme using advanced coding and modulations, thus the data rate is increased from 2 Mbps to 11 Mbps. Besides, 802.11b is backward compatible with 802.11. The data rate gap between 802.11a and 802.11b motivates further enhancements on PHY schemes in the 2.4 GHz band. In 2003, 802.11g was approved, which is backward compatible with both 802.11 and 802.11b. 802.11g still operates in the 2.4 GHz ISM band, and it also adopts OFDM into the PHY layer. The maximum data rate of 802.11g is also 54 Mbps. Currently, Task Group n is working on the advanced PHY techniques for future wireless LAN. It aims to support data rate up to 100 Mbps excluding MAC overhead. Many state-of-the-art technologies in communications, e.g.,MIMO, space-time signal processing, LDPC coding, are very likely to be finalized in the published standard. Future wireless LAN also requires QoS be provided for the diverse multimedia applications. This motivates Task Group e to develop QoS mechanisms over wireless LAN. The schemes being discussed include admission control, various contention window sizes for different applications, arbitration interframe space, etc. Table 8.1 summarizes the key parameters of each 802.11 version.

Recently, the Wireless Ethernet Compatibility Alliance (WECA) has proposed their certification program for 802.11 products. Any product passed their interoperable test program can be named as WiFi (wireless fidelity) products. As a consequence, 802.11 is sometimes referred to as WiFi.

8.1.2   802.11 network architecture

802.11 defines three types of network architecture: Independent basic service set (IBSS), basic service set (BSS) and extended service set (ESS). The wireless terminal in the 802.11 network is termed "station".

Standard	Frequency Band (GHz)	Published Time	PHY Technolo-gies	Maximum Rate (Mpbs)
802.11	2.4	1997	DSSS, FHSS and Infrared	2
802.11a	5	1999	OFDM	54
802.11b	2.4	1999	HR-DSSS	11
802.11g	2.4	2003	DSSS, OFDM	54
802.11n	Not yet	Not yet	OFDM, MIMO, LDPC, turbo,space-time codes	100 (ex-cluding MAC overhead)

Table 8.1: Current 802.11 standards comparison

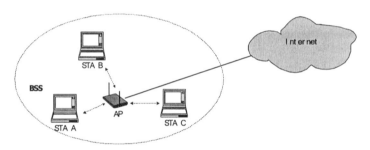

Figure 8.2: An example of Basic Sevice Set (BSS)

- BSS is also referred to as infrastructure network. In a BSS network, stations do not communicate with each other directly, but they communicate with a special terminal called access point (AP). The AP forwards the frames from the originating station to the destination station.

Example 45 *An example of an infrastructure network is shown in Figure 8.2 where stations communicate with the AP directly. The AP is usually connected to the backbone wired network. Therefore, stations in the BSS network have access to the backbone network with the aid of the AP. Frames destined to the backbone network are distinguished by the AP and forwarded to the corresponding backbone routers.*

- IBSS is also referred to as Ad Hoc network. In an IBSS network, stations communicate with each other directly and there is no access point

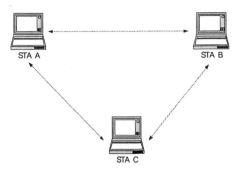

Figure 8.3: An example of Independent Basic Service Set (IBSS)

participating in the coordination. Therefore stations in the network are self-organized.

Example 46 *An IBSS network is shown in Figure 8.3 which contains three stations communicating directly with each other. Figure 8.3 can represent the scenario of a small conference meeting where stations share their information during the conference period.*

An IBSS network is usually within short range and contains a small number of devices. Besides, stations in IBSS cannot access the backbone network, as there is no AP forwarding their packets.

- An ESS is formed when several APs are connected. The component connecting the APs is called "distribution system" (DS). The DS is responsible for distributing the frames from the originating AP (i.e., the AP that communicates with the originating station) to the destination AP (i.e., the AP that communicates with the destination station). For frames directed to/from the backbone network, the DS is responsible for distributing these frames to the corresponding routers/APs.

Example 47 *An example of an ESS network is shown in Figure 8.4. The illustrated ESS contains three BSSs and the corresponding APs are connected by a hub. The hub in this case plays the role of a DS.*

Note that a DS is not necessarily of wired configuration. APs can also be connected through the wireless medium using the so called "wireless bridge" configurations.

From protocol stack point of view, the 802.11 family is similar to other 802 MAC and PHY specifications such as 802.3 and 802.5. However it faces

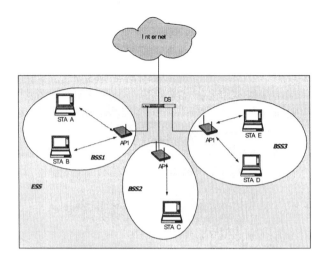

Figure 8.4: An example of Extended Service Set (ESS)

new challenges in the wireless medium. The path loss, shadowing and fading in the wireless environment give rise to high error rate in the communication link. Besides, unlike wired devices that can transmit and receive simultaneously, wireless devices are usually half duplex and cannot transmit and receive at the same time. These differences pose new challenges for the MAC and PHY layers in wireless LAN. In the following, we discuss the MAC and PHY technologies in 802.11 networks.

8.1.3   MAC layer technologies

Hidden node problem and collision avoidance

Generally speaking, 802.11 uses a contention based medium access scheme similar to Ethernet (802.3). However, the CSMA/CD scheme defined in 802.3 has its limitations when applied to the wireless environment. First of all, in Ethernet, the carrier sensing and collision detection are implemented by monitoring the wire's signal level by each station. A collision is assumed when the station detects an unusually high signal level (caused by simultaneous transmission of multiple stations). Secondly, the transmitted signal in Ethernet is assumed to reach all the stations connected on the wire. This is because the wired link has very low error rates and is regarded as very stable. However these two properties of Ethernet do not hold in wireless LAN. First of all, the path loss, shadowing and fading make the wireless signal level vary dramatically. Therefore, it is unlikely to justify whether the medium is busy or whether there is a collision by purely monitoring the received signal level. Secondly, because of the high error rate presented in the wireless link, the signal transmitted by one station is not guaranteed to be received correctly by another station. Furthermore, the

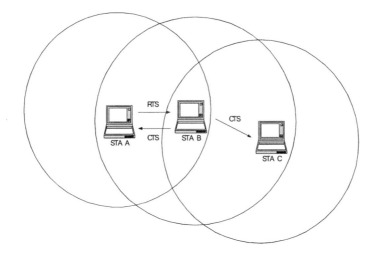

Figure 8.5: An example of the hidden node scenario

transmitted signal may not be even detected by another station, leading to the so-called "hidden node" scenario in wireless LAN.

Example 48 *Figure 8.5 shows an example of the "hidden node" scenario. There are three wireless devices– stations A, B and C. The circle centered at each station represents their transmission range. Stations outside of the circle cannot receive the signal sent by the station. In this example, A and B are within each other's range. B and C are within each other's range. However due to some reasons such as long distance, A and C cannot communicate with each other. In this scenario, when A is transmitting, C still regards the medium as free as it does not detect the signal transmitted from A. As a result, C may start transmitting while A has not finished, leading to collisions. Despite of the retransmission efforts, C will always interfere with A's transmission. In this case, C is regarded as a "hidden node" to A. Similarly, A is also a hidden node to C.*

To prevent collisions, 802.11 MAC layer uses a collision avoidance (CA) scheme. The transmitting station first sends a short message called "request to send" (RTS) when it has data to send. The receiver then responds with a short message called "clear to send" (CTS). These short messages contain information of how long the wireless medium needs to be reserved for their communications. All the stations hearing either of the two messages should set the medium state as busy and hold their transmission during this period. Therefore, the hidden nodes are silenced by the RTS and/or CTS. In the example in Figure 8.5, although the RTS message sent from A does not reach C, the CTS message sent from B is received by C. Therefore station C shall hold its transmission until A and B finish their communications. Note that in order to save the signaling overhead, 802.11 also has an option of sending data frames without exchanging RTS/CTS if the data frame is shorter than a threshold. This threshold is pre-set by the 802.11 network administrator.

In addition to the RTS and CTS exchange, data frames also indicate the medium reservation period. Moreover, acknowledgment frames are sent from the receiver to the sender when data frames are received correctly. These acknowledgment frames also indicate the amount of time that the medium will be reserved for the rest of the communication. As a result, the RTS/CTS signaling together with the data/acknowledgment medium reservation scheme prevent the hidden node problem in 802.11 networks.

Carrier sensing scheme

As mentioned earlier, the carrier sensing scheme used in Ethernet cannot apply to the wireless network due to the dramatic signal level variations. Instead, the carrier sensing in 802.11 is realized by physical layer's carrier sensing/clear channel assessment (CA/CCA) procedure together with the "network allocation vector" (NAV) in the MAC layer. In the physical layer, the carrier sensing is realized by a CA/CCA procedure using schemes such as signal detection and energy detection. While in the MAC layer, the carrier sensing is realized by monitoring NAV. NAV is a value stored in the MAC layer in each station. It indicates how much time the medium will still be busy. This value is updated by each station when it detects a larger NAV value in the received frame even if the frame is not addressed to this station. The NAV value counts down with time. When NAV reaches zero, the wireless medium is deemed as free from the MAC layer perspective. Otherwise, the station should hold its transmission until NAV goes down to 0. For example, when a station sends a RTS to the medium, it shall calculate the amount of time that is needed to transmit the responding CTS and the following data frames plus the pre-determined interframe periods. The station then sets this time in the frame header when the RTS frame is transmitted. All the stations receiving the RTS shall update their NAV values and count down with time.

Interframe spacing

By the time the station's NAV reaches zero and the PHY also reports an idle medium, the station must wait for an interframe space (IFS) before it starts the backoff timer to contend for the medium, which is shown in Figure 8.6. IFS plays an important role in regulating medium access. The type of frame that is to be sent determines the IFS that the station must wait in addition to the backoff timer. It is obvious that frames with shorter IFS have higher priority gaining the medium than frames with longer IFS. 802.11 defines four types of IFS for different frame types. In the following, we describe the four types of IFS in ascending order.

- Short interframe space (SIFS)

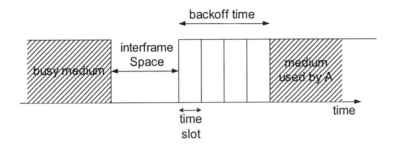

Figure 8.6: Interframe space and contention window

SIFS is the shortest IFS defined in 802.11. It is applied to the frames with the highest priority. For instance, when a station receives RTS, CTS is transmitted after SIFS period of time in order to respond to the originating station as soon as possible. Acknowledgments of correctly received frames also use SIFS to feed back the information to the sender quickly. When a data block exceeds the maximum frame length, fragmentation is required. The fragmented frames belonging to the same block use SIFS to facilitate fast assembly of the original data block without interruption from other stations. These frames have the highest priority and are transmitted after SIFS amount of time once the medium becomes idle, preempting over other types of frames.

- PCF interframe space (PIFS)

PIFS is used in contention free operation mode. In addition to contention based access method, 802.11 also defines a contention free operation mode. Contention free operation is regulated by a point coordination function (PCF). During contention free period, the AP polls each station and allows stations to transmit alternatively. The frames transmitted during contention free period use PIFS. It is shorter than the interframe space operated in contention-based period. Therefore, frames transmitted in contention free mode have higher priority gaining the medium than the regular contention-based frames.

- DCF inter-frame space (DIFS)

DIFS is the most commonly used IFS in 802.11 networks. The CSMA/CA and random backoff schemes are regulated by the distributed coordination function (DCF) in the MAC layer and DIFS is the default IFS coordinated by DCF.

- Extended inter-frame space (EIFS)

EIFS is used by the DCF whenever a frame is not received correctly with a valid CRC. When such a frame is determined to be erroneous, the next transmission attempt shall use EIFS instead of DIFS. Once the station receives a correct frame with a valid CRC, the following frames shall start using the regular DIFS.

EFIS is designed to provide enough time for another station to acknowledge, to this station, an incorrectly received frame before this station commence transmission. Reception of an error-free frame re-synchronizes the station with the actual medium idle/busy state; thus, EIFS is terminated, and the DIFS is used instead.

Random backoff contention scheme

After the medium has been sensed as idle for the corresponding interframe space period, the station shall contend for the wireless medium through DCF, if it has data to send.

Basically DCF regulates the CSMA/CA and random backoff contention procedure. The station randomly selects a value between 0 and its current contention window $(CW_{current})$. This value corresponds to the number of time slots the station must wait before it may transmit. For example, if station A has a contention window $CW_{current} = 15$, and it randomly selects a value between 0 and 15, e.g., 4, then the station must wait for additional 4 time slots before it may transmit. During this period, the station keeps sensing the medium. If the medium is sensed as free during these time slots, then A starts transmission as shown in Figure 8.6. However, if station B selects a shorter backoff value, e.g., 2 time slots, and starts transmission before A, then A should hold its data and update its NAV value after it receives frames from the medium. Once B completes its transmission, A resumes its count down procedure and starts transmission when the remaining 2 time slots elapse, which is shown in Figure 8.7.

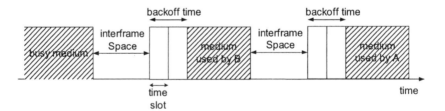

Figure 8.7: Random backoff and medium contention of two users

$CW_{current}$ is of the following form:

$$CW_{current} = 2^n - 1,$$

where n is an integer. The value of n is restricted by the inequality

$$CW_{min} \le CW_{current} \le CW_{max},$$

where CW_{min} and CW_{max} are pre-set values. For example, if $CW_{min} = 15$ and $CW_{max} = 127$. Then n can only take values from 4 to 7. Initially, $CW_{current}$ is

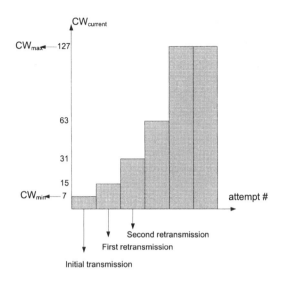

Figure 8.8: Exponential increase of CW

set to CW_{min}. $CW_{current}$ is also reset to CW_{min} when a frame is transmitted successfully. Every time the transmitted frame does not receive a corresponding acknowledgment or the responding message within a defined time, n increases by 1, i.e., $CW_{current}$ almost doubles as shown in Figure 8.8, until it reaches the allowed maximum value. In the following transmissions, the station shall use the new $CW_{current}$ value. In the meantime a retry count is increased by 1. The retry count also has an allowable maximum value. An error is reported to the upper layers when the maximum retry count is reached, and the frame is discarded.

Framing format

The information delivered from the MAC layer of one station to the peer layer of another station is encapsulated into a defined format. All MAC frames are generated following the format shown in Figure 8.9. All frames are made up of three parts: a frame header, a frame body and a frame check sequence (FCS). The subfields are described as follows:

- Frame Control field is comprised of the following subfields as shown Figure 8.10: Protocol Version, Type, Subtype, To DS, From DS, More Fragments (More Frag), Retry, Power Management (Power Mgmt), More Data, Wired Equivalent Privacy (WEP) and Order.

    - Protocol Version is a 2-bit field indicating which version of 802.11 MAC is functioning. At present, only one 802.11 MAC has been

Figure 8.9: 802.11 MAC frame format

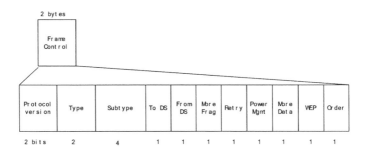

Figure 8.10: Frame control field

published, so this value shall be set to 0 and all other values are reserved.

- Type and Subtype fields together define the unique identity of the frame. There are three types of frames defined in the standard:

  Control frames, which are used to gain access of the wireless medium.

  Management frames, which are used to exchange management information. These frames are transmitted as data frames, but not passed to the upper layer.

  Data frames are used for data transmission. These frames are passed to upper layers.

  For each frame type, the subtype field further specifies the exact function of the frame. The valid type and subtype combination is shown in Table. 8.2. Other combinations are reserved.

- To DS/From DS field indicates whether the frame is directed to/coming from DS.

- More Fragments field indicates whether this frame is a fragment of a large MAC SDU.

- Retry field indicates whether the frame is a retransmission frame.

- Power Management field indicates whether the station is in power save mode.

Type Value	Type Description	Subtype Value	Subtype Description
00	Management	0000	Association request
00	Management	0001	Association response
00	Management	0010	Reassociation request
00	Management	0011	Reassociation response
00	Management	0100	Probe request
00	Management	0101	Probe response
00	Management	1000	Beacon
00	Management	1001	ATIM
00	Management	1010	Disassociation
00	Management	1011	Authentication
00	Management	1100	Deauthentication
01	Control	1010	Power save (PS)-Poll
01	Control	1011	RTS
01	Control	1100	CTS
01	Control	1101	Ack
01	Control	1110	Contention free (CF)-End
01	Control	1111	CF-End+CF-Ack
10	Data	0000	Data
10	Data	0001	Data+CF-Ack
10	Data	0010	Data+CF-Poll
10	Data	0011	Data+CF-Ack+CF-Poll
10	Data	0100	Null function (no data)
10	Data	0101	CF-Ack (no data)
10	Data	0110	CF-Poll (no data)
10	Data	0111	CF-Ack+CF-Poll (no data)

Table 8.2: Frame Type and Subtype combinations

- More Data field indicates whether there is more data to send to/from the station.
- WEP field indicates whether the Wired Equivalent Privacy (WEP) encryption method is used in the frame body.
- Order field indicates whether the station is reordering the MAC SDUs. The reordering is to adjust the delivery order of broadcast/multicast MAC SDUs relative to unicast SDUs. This service is designed to improve the likelihood of successful delivery.

• Duration ID, in most cases, is set to be the amount of time that the medium is reserved for the station. In certain frame types, this field contains the station's Association ID.

• Address 1, 2, 3 and 4 fields represent BSS Identification, source address, destination address, transmitting station address, and receiving station

4 bits   12 bits

Figure 8.11: Sequence control field

address, respectively. Certain frame types may not contain some of the address fields.

• Sequence Control field is comprised of two parts as shown in Figure 8.11

• The Sequence Number field is a 12-bit field indicating the sequence number of an MSDU or MAC management PDU (MMPDU). If the MSDU or MMPDU is larger than the maximum frame length, it shall be fragmented into small segments to fit into the frame format. The Fragment Number then indicates the fragment number belonging to the current MSDU or MMPDU.

• Frame Body field is a variable length field and contains information directed to the peer MAC layer. The detailed information is dependent on the specific frame type and subtype.

• FCS is a 32-bit CRC calculated over all the fields in the MAC header and the Frame Body.

Example 49 *An example of MAC frame exchanges between a station and an AP is shown in Figure 8.12. The message exchange in this example is described as follows:*

1. *After the station powers up, a Probe Request is first transmitted to detect available APs around it. Probe Request contains the station's capability information such as supportable rate.*

2. *If the Probe Request is received successfully by an AP, the AP shall respond with a Probe Response to inform the AP's information such as timing, beacon signal, supportable rate and PHY layer parameters.*

3. *If the station decides to connect to the AP, an Authentication procedure is invoked to determine the validity of the station on this network. This is a two-way Authentication procedure.*

4. *After the station passes the authentication procedure, it sends an Association Request to associate itself to the current AP. Association Request contains the stations's information such as power save mode parameters as well as supportable rate information.*

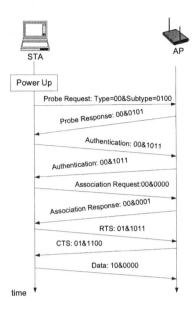

Figure 8.12: An example of MAC frame exchanges between an station and an AP

5. *If the AP accepts the station, an Association Response is sent to the station with the assigned Association ID as well as the supportable rates by the AP.*

6. *If the station has data to send, it first contends the medium with RTS/CTS.*

7. *If the station gains control of the medium, the station starts data transmission.*

8.1.4 Physical layer technologies

Up till now, IEEE has published five physical layer options for 802.11 networks: FHSS and DSSS (802.11) in the 2.4 GHz band, HR-DSSS (802.11b) in the 2.4 GHz band, Further Higher Data Rate Physical Layer Extension in the 2.4 GHz band and OFDM (802.11a) in the 5 GHz band. Please refer to Table 8.1 for their comparisons. Currently, 802.11 Task Group n is working on the future wireless LAN physical layer specifications. Many updated communication technologies, e.g., MIMO, LDPC coding, space-time code, are to be finalized into the standard.

Generally speaking, the 802.11 physical layer is comprised of two sublayers: physical layer convergence procedure (PLCP) and physical medium dependent (PMD) as shown in Figure 8.13. However, the detailed specifications of PLCP and PMD are dependent on the individual PHY option. PLCP defines the

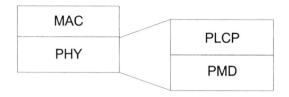

Figure 8.13: Physical layer procedures

method of mapping the MAC PDU (MPDU) onto the framing format that is suitable for the associated PMD transmissions/reception. PLCP thus provides a means to minimize the interaction of the MAC layer with the actual physical operation. PMD defines the characteristics and method of transmitting and receiving data between two stations. It provides the actual means of transmitting and receiving data using the specific PHY scheme. Since this book is centered around OFDM technologies, we focus on 802.11a, which uses OFDM in the physical layer.

In 802.11a, when an MPDU is handed from the MAC layer to the physical layer, PLCP first processes the MPDU to form the physical layer PDU (PPDU) described as follows. A PLCP preamble and a PLCP header are added in front of the MPDU. A Tail field and a Pad field are appended after the MPDU. The PLCP preamble and header are used to aid the demodulation/decoding and delivery of the MPDU at the receiver side. On the other direction, when a PLCP receives a PPDU, it processes the PLCP header to decode the embedded MPDU. The decoded MPDU is then delivered to the MAC layer.

Each PPDU is made up with three parts: Preamble, SIGNAL and DATA. The SIGNAL field is also a part of the PLCP header as shown in Figure 8.14.

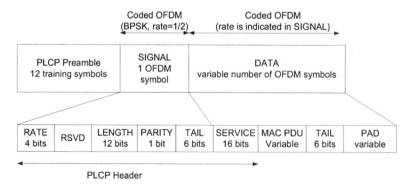

Figure 8.14: 802.11a PLCP frame format

• PLCP Preamble

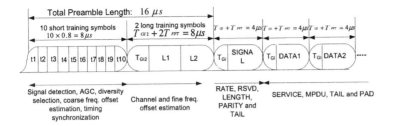

Figure 8.15: 802.11a OFDM training sequence structure

PLCP Preamble contains 10 short training symbols and two long training symbols. The short symbols are used for signal detection, automatic gain control, diversity selection, coarse frequency offset estimation and timing synchronization. The long symbols are used for channel and fine frequency offset estimation. The structure of the training symbols is shown in Figure 8.15.

- The related parameters in Figure 8.15 are defined in Table. 8.3

Parameter	Value
Number of data subcarrier	48
Number of pilot subcarrier	4
Subcarrier frequency spacing ΔF	$0.3125 \ (20M/64)$
1IFFT/FFT Period T_{FFT}	$3.2\mu s \ (\frac{1}{\Delta F})$
PLCP Preamble duration $T_{preamble}$	$16\mu s$
Guard duration T_{GI}	$0.8\mu s \ (\frac{T_{FFT}}{4})$
Long training symbol guard duration T_{GI2}	$1.6\mu s \ (\frac{T_{FFT}}{2})$
Symbol interval T_{sym}	$4\mu s \ (T_{GI} + T_{FFT})$

Table 8.3: Timing dependent parameters

One OFDM symbol is transmitted across 52 subcarriers. Four of them are pilots and the remaining 48 carry data. A general OFDM symbol has a 0.8 μs guard interval followed by 3.2 μs data duration. For the long training symbols, the guard interval is 1.6 μs. The frequency domain of the short training sequence is shown in Table 8.4: the short sequence is transmitted on 12 out the 52 subcarriers. In order to normalize the average transmitted power on each subcarrier, a normalizing factor multiplies the short training sequence, i.e.,

$$K_{norm} \times \left(12 \times (1^2 + 1^2)/52\right) = 1.$$

#	$\mathrm{Re}(\cdot)$ $\times\sqrt{\frac{13}{6}}$	$\mathrm{Im}(\cdot)$ $\times\sqrt{\frac{13}{6}}$	#	$\mathrm{Re}(\cdot)$ $\times\sqrt{\frac{13}{6}}$	$\mathrm{Im}(\cdot)$ $\times\sqrt{\frac{13}{6}}$	#	$\mathrm{Re}(\cdot)$ $\times\sqrt{\frac{13}{6}}$	$\mathrm{Im}(\cdot)$ $\times\sqrt{\frac{13}{6}}$
−32	0	0	−10	0	0	12	1	$-j$
−31	0	0	−9	0	0	13	0	0
−30	0	0	−8	−1	$-j$	14	0	0
−29	0	0	−7	0	0	15	0	0
−28	0	0	−6	0	0	16	−1	$-j$
−27	0	0	−5	0	0	17	0	0
−26	0	0	−4	1	$-j$	18	0	0
−25	0	0	−3	0	0	19	0	0
−24	1	j	−2	0	0	20	−1	j
−23	0	0	−1	0	0	21	0	0
−22	0	0	0	0	0	22	0	0
−21	0	0	1	0	0	23	0	0
−20	−1	j	2	0	0	24	1	j
−19	0	0	3	0	0	25	0	0
−18	0	0	4	1	$-j$	26	0	0
−17	0	0	5	0	0	27	0	0
−16	−1	$-j$	6	0	0	28	0	0
−15	0	0	7	0	0	29	0	0
−14	0	0	8	−1	$-j$	30	0	0
−13	0	0	9	0	0	31	0	0
−12	1	$-j$	10	0	0	−	−	−
−11	0	0	11	0	0	−	−	−

Table 8.4: Short training sequence represented in frequency domain

As a result, $K_{norm} = \sqrt{13/6}$.

The long training sequence is transmitted across 53 subcarriers, including a zero value at dc. The frequency domain representation of the long training sequence is shown in Table 8.5.

- SIGNAL field

  SIGNAL field is comprised of RATE, RSVD, LENGTH, PARITY AND SIGNAL TAIL fields as shown in Figure 8.14.

  - RATE field indicates the current data rate. Tables 8.6 shows the encoding of each supportable data rate. The rate-dependent parameters are listed in Table 8.7.

    The data rate is calculated as the number of data bits per OFDM symbol divided by one OFDM symbol interval . For example, 6Mbps is calculated as 24 bits/$4\mu s$ = 6 Mbps.

  - LENGTH field is the number of bits in the MPDU.

#	$\text{Re}(\cdot)$ $\times\sqrt{\frac{13}{6}}$	$\text{Im}(\cdot)$ $\times\sqrt{\frac{13}{6}}$	#	$\text{Re}(\cdot)$ $\times\sqrt{\frac{13}{6}}$	$\text{Im}(\cdot)$ $\times\sqrt{\frac{13}{6}}$	#	$\text{Re}(\cdot)$ $\times\sqrt{\frac{13}{6}}$	$\text{Im}(\cdot)$ $\times\sqrt{\frac{13}{6}}$
-32	0	0	-10	-1	0	12	-1	0
-31	0	0	-9	1	0	13	-1	0
-30	0	0	-8	1	0	14	-1	0
-29	0	0	-7	-1	0	15	1	0
-28	0	0	-6	1	0	16	1	0
-27	0	0	-5	-1	0	17	-1	0
-26	1	0	-4	1	0	18	-1	0
-25	1	0	-3	1	0	19	1	0
-24	-1	0	-2	1	0	20	-1	0
-23	-1	0	-1	1	0	21	1	0
-22	1	0	0	0	0	22	-1	0
-21	1	0	1	1	0	23	1	0
-20	-1	0	2	-1	0	24	1	0
-19	1	0	3	-1	0	25	1	0
-18	-1	0	4	1	0	26	1	0
-17	1	0	5	1	0	27	1	0
-16	1	0	6	-1	0	28	1	0
-15	1	0	7	1	0	29	1	0
-14	1	0	8	-1	0	30	1	0
-13	1	0	9	1	0	31	1	0
-12	1	0	10	-1	0	-	-	-
-11	-1	0	11	-1	0	-	-	-

Table 8.5: Long training sequence represented in frequency domain

- RSVD bit is reserved for future use
- PARITY is an even parity bit for the first 16 bits
- SIGNAL TAIL shall be set to all zeros.

- DATA

  DATA field is comprised of SERVICE, MPDU, TAIL and PAD fields as shown in Figure 8.14.

  - SERVICE field is also the last field in the PLCP header. The first 6 bits of SERVICE are set to zero to synchronize with the descrambler at the receiver side. The rest of the SERVICE bits are reserved for future use and are also set to zero.
  - MPDU contains the actual information intended to transmit/receive by the stations. It contains the MAC frames constructed according to MAC framing format.
  - TAIL bits are all zeros in order to initialize the convolutional encoder.

Data Rate (Mbps)	Bits
6	1101
9	1111
12	0101
18	0111
24	1001
36	1011
48	0001
54	0011

Table 8.6: SIGNAL field content

Data Rate (Mbps)	Modulation	Coding Rate	Coded Bits per Subcarrier	Coded Bits per OFDM Symbol	Data Bits per OFDM Symbol
6	BPSK	1/2	1	48	24
9	BPSK	3/4	1	48	36
12	QPSK	1/2	2	96	48
18	QPSK	3/4	2	96	72
24	16QAM	1/2	4	192	96
36	16QAM	3/4	4	192	144
48	64QAM	2/3	6	288	192
54	64QAM	2/4	6	288	216

Table 8.7: Rate dependent parameters

- PAD field contains a variable number of zero bits. It is used to pad the DATA filed into an integral number of OFDM symbols.

• The DATA field including SERVICE, PSDU, TAIL and PAD shall be scrambled with a length-127 frame-synchronous scrambler shown in Figure 8.16. The same scrambler is used to scramble the transmitted bits and descramble the received bits.

The scrambled bits are then passed into the convolutional encoder shown in Figure 8.17. The coded bits are interleaved and mapped to modulation symbols using BPSK, QPSK, 16-QAM or 64-QAM depending on the rate requested. IFFT/FFT is then implemented and cyclic prefix is added.

Note that in order to achieve the same average power for all modulation types, a normalizing factor multiplies each symbol. The normalizing factor depends on the modulation type and is shown in Table 8.8.

• Let us take BSPK and 16-QAM as an example of calculating the normalizing factor. The modulated signals for BPSK and 16-QAM are shown in Figure 8.18.

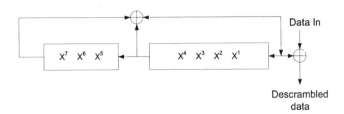

Figure 8.16: 802.11a DATA scrambler/descrambler structure

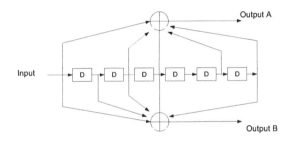

Figure 8.17: 802.11a convolutional encoder

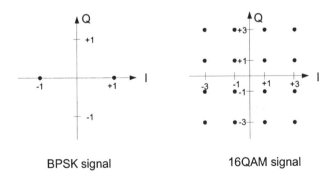

Figure 8.18: QPSK and 16-QAM modulated signals

Modulation	Normalizing Factor K_{MOD}
BPSK	1
QPSK	$1/\sqrt{2}$
16-QAM	$1/\sqrt{10}$
64-QAM	$1/\sqrt{42}$

Table 8.8: Modulation-dependent normalizing factor

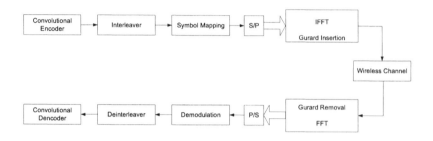

Figure 8.19: 802.11a PHY baseband transmitter and receiver blocks

The average received power for the BPSK signal is calculated as

$$P_{BPSK} = K_{BPSK} \times \frac{1}{2} \left(1^2 + 1^2 \right). \tag{8.1}$$

The average received power for the 16-QAM signal is calculated as

$$P_{16-QAM} = K_{16-QAM} \times \frac{1}{16} \left(\begin{array}{c} 4 \times \left(1^2 + 1^2\right) + 4 \times \left(1^2 + 3^2\right) \\ +4 \times \left(3^2 + 1^2\right) + 4 \times \left(3^2 + 3^2\right) \end{array} \right). \tag{8.2}$$

In order for (8.1) and (8.2) to be equal, the normalizing factor should have the following relationship:

$$K_{16-QAM} = \frac{K_{BPSK}}{\sqrt{10}}.$$

Finally, we summarize the 802.11a baseband transmitter and receiver in Figure 8.19.

8.2 IEEE 802.16e and Mobile WiMAX

The IEEE 802.16e is a suite of broadband wireless technologies that are complementary to IEEE 802.11 WiFi. In particular, the IEEE 802.16e standard

defines the broadband Wireless MAN (metropolitan area network) air interface specification that can provide services to many more users at much longer
distances. According to the IEEE -SA Project Authorization Request (PAR)
802.16e, the IEEE 802.16 Task Group e (Mobile WirelessMAN) develops "physical and medium access control layers for fixed and mobile operation in licensed
bands" (URL: Http://www.ieee802.org/16/tge/). The project provides enhancements to IEEE 802.16-2004 to support subscriber stations moving at vehicular speeds and thereby specifies a system for combined fixed and mobile
broadband wireless access. Functions to support higher layer handoff between
base stations or sectors are specified. Operation is limited to licensed bands
suitable for fixed/mobile user below 6GHz.

The IEEE 802.16e is a rapidly evolving draft that incorporates a wide range
of the state-of-the-art technologies. There are three basic PHY modes in the
802.16e documents:

1. single-carrier

2. OFDM

3. OFDMA

The single-carrier mode and the OFDM mode are both inherited from the
IEEE standard 802.16d fixed broadband wireless access systems. Most of the
design and standardization activities within 802.16e are on the OFDMA mode,
which provides the only scalable scheme for wide-area mobile networks for reasons stated in Section 5. This appendix, therefore, only describes the *OFDMA*-
based physical (PHY) layer and medium access control (MAC) layer within the
802.16e. Since the IEEE 802.16e is still in the final draft stage at the time of
this manuscript, some configurations/parameters are subject to further modifications. Readers are referred to the actual standard for the final details.

While IEEE creates standards, they do not have a process for driving confor-
mance, compliance and interoperability. The IEEE 802.16 is closely associated
with WiMAX, which stands for "Wireless (Wi) Microwave Access (MA)". The
WiMAX Forum (the Worldwide Interoperability for Microwave Access Forum)
is a non-profit corporation formed by equipment and component suppliers to
promote the adoption of IEEE 802.16 compliant equipment by operators of
broadband wireless access systems (URL: Http://www.wimaxforum.org). In
particular, fixed WiMAX is a standard-based technology enabling the delivery
of last mile wireless broadband access as an alternative to cable and DSL. It
is expected that the mobile WiMAX technology based on 802.16e will be in-
corporated in notebook computers and PDAs in 2006, allowing for urban areas
and cities to become "MetroZones" for portable and mobile outdoor broadband
wireless access.

Max subscriber through-put	1Mbps/uplink, 3Mbps/downlink
Max sector throughput (10Mhz band)	18Mbps/downlink; 6Mbps/uplink
Frequency reuse	1
Mobility	up to 120 Km/hour
Handoff	under 150ms
Service coverage	Macro (1Km), Micro (400m), Pico (100m)
Roaming	seamless roaming with cellular and WLAN
QoS offering	unsolicited grant service; real-time; non real-time; best-effort
Uplink/Downlink ratio	software adjustable

Table 8.9: Mobile WiMAX system highlights

8.2.1 Overview

Solving the last mile connectivity problem to a backbone network (such as the internet) continues to be a challenge of fundamental importance for evolution of next generation wireless networks. The IEEE 802.16e is positioned to provide broadband data link to pedestrian and mobile terminals within a 1-3 mile radius. The 802.16e comprises of a large variety of configurations, capable of supporting both TDD and FDD based networks with bandwidth ranging from 1.25MHz to 20Mhz. Table 8.9 highlights some of system features with regard to user experience and network performance.

Given its capabilities, the target applications of 802.16e are mostly data-oriented. In particular, the 802.16e access solution provides a unified network architecture to support heterogeneous traffic with different QoS requirements. Table 8.10 lists some exemplary applications of mobile WiMAX.

8.2.2 Physical layer technologies

In the OFDMA mode, the channel bandwidth is partitioned into subchannels that can be dynamically allocated to the mobile subscriber stations (MSS). Depending on the available bandwidth, at least one of the FFT sizes (2048, 1024, 512, 128) is supported. The FFT size may be dynamically detected by the MSS through scanning and searching of the downlink (DL) signals when performing initial network entry.

QoS Class	Data Type	Application
unsolicited grant service	periodic interval, fixed-sized packet; real time	T1/E1; VoIP with silience suppression
unsolicited grant service	periodic interval; variable-sized packet; real time data stream	video telephony; interactive video game; VoD/AoD
real time polling service	variable-sized packet; delay-tolerant data stream; minimum data rate	high speed file transfer; MMS; Web browsing
best effort service	no minimum service level	FTP, WWW, E-mail

Table 8.10: Mobile WiMAX applications

Many of the techniques described in the previous chapters are included in the OFDMA mode, thanks to the contributions from its members. The salient features of the IEEE 802.16e PHY are

- Robust data subchannels for mobility and aggressive frequency reuse

  - frequency diversity for high mobility
  - subcarrier grouping for inter-sector and inter-cell interference mitigation
  - frequency reuse of 1-3 in a multi-sector, multi-cell configuration

- Advanced modulation and coding (AMC) subchannels for multiuser and network diversity exploitation

  - dynamic subchannel allocation (intracell and potentially intercell)
  - channel quality information feedback and uplink channel sounding
  - adaptive coded modulation for maximum throughput

- Scalable OFDMA [7]

  - subcarrier spacing independent of bandwidth to provide both Doppler and multipath immunity
  - the number of used subcarriers (and FFT size) scales with bandwidth/FFT size
  - fixed smallest unit of bandwidth allocation

- Flexible subchannel structures and zone switching between diversity and AMC modes

- Channel coding in combination of hybrid automatic repeat request (H-ARQ)

- MIMO and advanced antenna systems (AAS)

parameter	Values			
multiple-access	OFDMA			
bandwidth [MHz]	1.25	5	10	20
FFT size	128	512	1024	2048
tone spacing [KHz]	11.16			
cyclic prefix overhead	1/8			
OFDMA symbol duration [us]	100.8			
number of symbols in a frame	same, depending on the frame length			

Table 8.11: Scalable OFDMA configurations

Basic functions and system parameters

The PHY layer of the 802.16e standard defines specifications regarding

- frame structures, including uplink and downlink frames and optional antenna array (AAS) frames

- map message fields and uplink (UL) and downlink (DL) information element (IE) formats

- the OFDMA subcarrier allocation, including downlink preamble, data subchannel and pilot structures, uplink data subchannel and channel sounding.

- OFDMA ranging signals

- antenna array operations, including 2x, 3x, and 4x space-time codes and antenna beamforming mechanisms.

- channel coding and modulation, hybrid ARQ mechanisms

- power control

In scalable OFDMA, system parameters such as (1) the OFDM symbol duration, (2) the frame size, (3) the subcarrier tone spacing are harmonized. In particular, the scalable OFDMA eliminates many unnecessary bandwidth dependent configurations. Such scalability allows a much more simplified terminal modem realization. Table 8.11 highlights the key parameters of the scalable OFDMA system

Table 8.12 lists the uplink and downlink data rates in a 10MHz OFDMA system with 10ms frames and a 3:1 downlink/uplink ratio. The raw rates are calculated based on the used bandwidth (subcarriers) and modulation schemes, whereas the net rates exclude the pilot and cyclic prefix overhead. Note that under different configurations (e.g., frame sizes, uplink/downlink ratios, and subchannel modes), the data rate will vary. Also in many practical situations, it may be improbable to reach the uplink peak rate (e.g., 64-QAM) due to link budget constraints.

Data Rate (Mbps)	Raw	DL raw	UL raw	DL net	UL net
QPSK 1/2	8.89	6.67	2.22	5.68	1.84
QPSK 2/3	11.85	8.89	2.69	7.58	2.46
QPSK 3/4	13.33	10.00	3.33	8.52	2.76
16QAM 1/2	17.78	13.33	4.44	11.37	3.69
16QAM 2/3	23.70	17.78	5.93	15.16	4.92
16QAM 3/4	26.67	20.00	6.67	17.05	5.52
64QAM 1/2	26.67	20.00	6.67	17.05	5.52
64QAM 2/3	35.56	26.67	8.89	22.73	7.37
64QAM 3/4	40.00	30.00	10.00	25.57	8.29
64QAM 5/6	44.44	33.33	11.11	8.29	9.22

Table 8.12: Row and net data rate of 10MHz TDD OFDMA system

Frame structure   The scalable OFDMA supports both TDD and FDD frame structures, with the frame size ranging from 2 msec to 20 msec. A representative TDD frame is depicted in Fig. 8.20.

Each frame is divided into four regions: DL (downlink transmission), TTG (transmit transition gap), UL (uplink transmission), and RTG (receive transition gap). The TTG and RTG provide guard periods against round trip delay in TDD operation as well as a ramping down period of the power amplifiers.

- The preamble is a PN sequence modulated on the preamble subcarrier set. The preamble carries the Cell ID and segment information for the MSS. In addition, the MSS can perform initial timing and carrier synchronization based on the preamble signals.

- The downlink subframe following the preamble shall always start in PUSC mode (defined in next section). Allocations subsequent to it shall use other modes as instructed.

- Following the preamble is the frame control header (FCH) that contains 48 bits modulated by QPSK with coding rate 1/2 and repetition coding of 4. The FCH contains the DL Frame Prefix (DLFP) to specify the burst profile and, in particular, the DLFP defines

  - the used subchannel bitmap
  - the ranging change
  - the coding format and repetition on the DL-MAP immediately following the FCH.
  - the length of the DL-MAP

- If needed, the DL-MAP follows the FCH to define the usage of the downlink intervals for a burst mode PHY. The basic DL-MAP specifies

- the connection identifier (CID) that represents the assignment of the IE to a broadcast, multicast, or unicast address

- Downlink interval usage code (DIUC) that defines the type of downlink burst type.

- OFDMA symbol offset at which the burst starts (in units of OFDMA symbols)

- subchannel offset - the lowest index subchannel used for carrying the burst

- power boost applied to the allocated data subcarriers.

- No. of OFDMA symbols - the number of OFDMA symbols that are used to carry the uplink burst

- the number of subchannels with subsequent indices

- repetition code used inside the allocated burst

- The UL-MAP immediately follows DL-MAP (if transmitted) or the FCH. The OFDMA UL-MAP IE defines uplink bandwidth allocations. The basic UL-MAP IE specifies

    - the connection identifier (CID) that represents the assignment of the IE to a broadcast, multicast, or unicast address

    - Uplink interval usage code (UIUC) that defines the type of uplink access and the burst type associated with the access.

    - OFDMA symbol offset at which the burst starts (in units of OFDMA symbols)

    - subchannel offset - the lowest index subchannel used for carrying the burst

    - No. of OFDMA symbols - the number of OFDMA symbols that are used to carry the uplink burst

    - the number of subchannels with subsequent indices

    - the duration, in units of OFDMA slots, of the allocation

    - repetition code used inside the allocated burst

Extended IE formats are indicated by DUIC=15 (UIUC=15) to allow specific structures (e.g., MIMO) in the subsequent allocations.

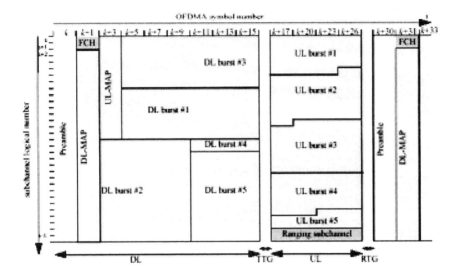

Figure 8.20: OFDMA TDD frame structure (source: IEEE 802.16 standard)

Data channel configurations

A basic subchannel in OFDMA always contains a total of 48 subcarriers out of one or several OFDM symbols. There are several ways to classify the OFDMA data subchannels:

- Downlink vs. uplink: there are three basic types of downlink subchannels (FUSC, PUSC and AMC) and two basic types of uplink subchannels (PUSC, AMC). The PUSC subchannel patterns for uplink and downlink are different.

- Diversity vs. adjacent: each diversity subchannel (i.e., FUSC, PUSC, optional FUSC - OFUSC, optional PUSC - OPUSC) consists of distributed subcarriers, whereas an adjacent subchannel (i.e., AMC) is comprised of consecutive subcarriers.

- Pre-pilot vs. post-pilot: except for FUSC, all subchannel configurations have the pilots and data subcarriers together in some fixed units (e.g., a cluster, a bin, or a tile). In FUSC, the subcarriers are divided into subchannels *after* the pilot tones are already in their positions.

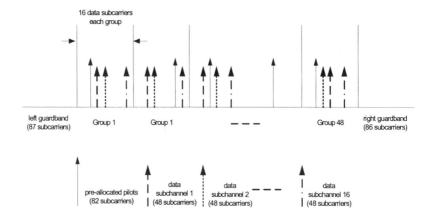

Figure 8.21: 1024-FFT OFDMA DL FUSC subcarrier configurations

Downlink fully used subchannelization (FUSC)

FUSC is a diversity DL channel configuration suitable for mobile or heavily interfered subscriber stations. Figure 8.21 illustrates the 1024-FFT, 10MHz bandwidth case where the DL FUSC contains

- 16 subchannels

- each subchannel has 48 subcarriers

- the 48 subcarriers within each subchannel are taken from 48 separate groups

- each group consists of 16 contiguous data subcarriers.

The exact allocation of the subcarriers into subchannels is done according to an equation called DL permutation formula. Table 8.13 summarizes the basic configuration of the FUSC.

Partially used subchannelization (PUSC)   The PUSC mode is mandatory for the first subframe in both downlink and uplink transmissions. Figure 8.22 illustrates the basic units in the PUSC subchannels.

- The data subcarriers in DL are partitioned into *clusters*, each of which contains a total of 28 subcarriers over 2 OFDM symbol periods. Clusters are then rearranged and partitioned into 6 non-overlapping *groups*, which

Bandwidth (MHz)	1.25	5	10	20
FFT size	128	512	1024	2048
# of guard subcarriers (left)	11	43	87	173
# of guard subcarriers (right)	11	43	86	172
# of data subcarriers	96	384	768	1536
# of pilot subcarriers	9	42	83	166
# of data subchannels (# of subcarriers per group)	2	8	16	32
# of subcarrers per subchannel (# of groups)	48	48	48	48

Table 8.13: FUSC channel configuration

can be allocated to different sectors or cells. Within each group, 48 sub-carriers are mapped into subchannels based on a permutation mechanism that minimizes the probability of hits between groups.

- Unlike DL, the basic unit in a UL PUSC subchannel is called a *tile*. The whole frequency band is partitioned into groups of contiguous tiles. Each subchannel consists of 6 distributed tiles, where each tile is chosen from different groups. Each tile spans 4 (or 3) subcarriers over 3 OFDM symbol periods. The 4×3 tile contains more pilots than the 3×3 and, therefore, is more robust. Each subchannel is comprised of 6 tiles, and the exact allocation of tiles into a subchannel is defined by the *uplink permutation formula*. As always, there are a total of 48 subcarriers in each subchannel.

- Tile-based subchannels (TUSC1 and TUSC2) are available in downlink as an option within an AAS zone.

Uplink and downlink advanced modulation and coding subchannels

(AMC)   In both the uplink and downlink AMC modes, the data subcarriers are partitioned into bands of consecutive subcarriers. Each band has 4 bins, which is a collection of 8+1=9 subcarriers. Depending on how the 6 bins ($6 \times 8 = 48$ subcarriers) are arranged into a subchannel, there are four types of AMC subchannels – see Figure 8.23 for illustration.

Channel information is critical in the effective operation of the AMC mode. To perform downlink AMC, uplink channel sounding is needed to enable the BS to determine the BS-to-terminal channel response under the assumption of TDD reciprocity. A dedicated Sounding Zone is allocated in which one or more

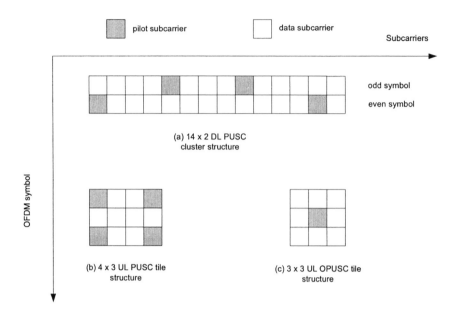

Figure 8.22: UL and DL PUSC structures

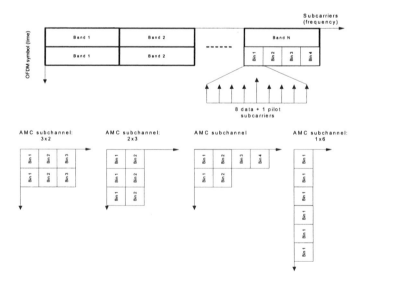

Figure 8.23: Bin arrangements in AMC subchannels

OFDMA symbol intervals in the UL frame are used by the MSS to transmit sounding signals.

From an uplink link budget perspective, the AMC mode 4 is most power efficient since it only utilizes $8+1 = 9$ subcarriers in each OFDM symbol (instead of $4 \times 6 = 24$ subcarriers/symbol in PUSC and $3 \times 6 = 18$ subcarriers/symbol in OPUSC). The power saving is therefore $3 \sim 4.26$ dB.

Modem and channel coding

BPSK, Gray-mapped QPSK, 16-QAM, 64-QAM are supported in 802.16e. The channel coding options in 802.12e include

- repetition coding (2x, 4x, and 6x) for control signals

- convolutional coding (CC) with incremental redundancy

- convolutional turbo coding (CTC) with incremental redundancy

- block turbo coding (BTC)

- low-density parity-check coding (LDPC)

An incremental redundancy based H-ARQ takes the puncture pattern into account, and for each retransmission the coded block is not the same. At the receiver, the received signals are de-punctuated according to their specific puncture pattern, then the combination is performed at bit matrix level.

Antenna diversity MIMO/space-time coding is optional for both uplink and downlink. Since the number of antenna elements at the MSS is mostly limited, only 2x STC is defined for uplink, whereas the dimension of downlink STC can go up to four dimensions.

The 2x Alamouti STC and spatial multiplexing (SM) can be used within the optional uplink PUSC and AMC zones to improve the system performance. The following matrices define the transmission format with the row index indicating antenna number and the column index indicating OFDM symbol time.

$$\mathbf{A} = \begin{bmatrix} s_i & -s_{i+1}^* \\ s_{i+1} & s_i^* \end{bmatrix}$$

$$\mathbf{B} = \begin{bmatrix} s_i \\ s_{i+1} \end{bmatrix}$$

The matrix $\mathbf{B}$ may also be used for two single antenna MSSs to share the same subchannel in an SDMA fashion.

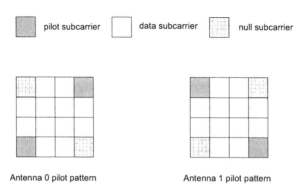

Figure 8.24: Uplink pilot patterns for 2-antenna PUSC

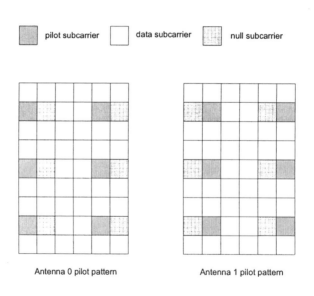

Figure 8.25: Uplink pilot patterns for 2-antenna AMC subchannels

Two optional zones with expanded tile/bin are defined for the uplink MIMO operations. Figures 8.24 and 8.25 show the pilot patterns for the optional PUSC and AMC zones, respectively.

2x, 3x, and 4x STC, including OSTBC and spatial multiplexing, are available for downlink MIMO. In addition, the STC outputs may be weighted (beamforming) before mapping onto transmission antennas (see Chapter 4 for discussion).

$$\mathbf{Z} = \mathbf{WX}$$

The transmission schemes corresponding to different antenna configurations are listed as following

- 2-antenna BTS downlink

$$\mathbf{A} = \begin{bmatrix} s_i & -s_{i+1}^* \\ s_{i+1} & s_i^* \end{bmatrix}$$

$$\mathbf{B} = \begin{bmatrix} s_i \\ s_{i+1} \end{bmatrix}$$

$$\mathbf{C} = \frac{1}{\sqrt{1+r^2}} \begin{bmatrix} s_i + jrs_{i+3} & rs_{i+1} + s_{i+2} \\ s_{i+1} - rs_{i+2} & jrs_i + s_{i+3} \end{bmatrix} \quad r = \frac{-1+\sqrt{5}}{2}$$

- 3-antenna BTS downlink

$$\mathbf{A}_1 = \begin{bmatrix} \tilde{s}_1 & -\tilde{s}_2^* & 0 & 0 \\ \tilde{s}_2 & \tilde{s}_1^* & \tilde{s}_3 & -\tilde{s}_4^* \\ 0 & 0 & \tilde{s}_4 & \tilde{s}_3^* \end{bmatrix}$$

$$\mathbf{A}_2 = \begin{bmatrix} \tilde{s}_1 & -\tilde{s}_2^* & \tilde{s}_3 & -\tilde{s}_4^* \\ \tilde{s}_2 & \tilde{s}_1^* & 0 & 0 \\ 0 & 0 & \tilde{s}_4 & \tilde{s}_3^* \end{bmatrix}$$

$$\mathbf{A}_3 = \begin{bmatrix} \tilde{s}_1 & -\tilde{s}_2^* & 0 & 0 \\ 0 & 0 & \tilde{s}_3 & -\tilde{s}_4^* \\ \tilde{s}_2 & \tilde{s}_1^* & \tilde{s}_4 & \tilde{s}_3^* \end{bmatrix}$$

$$\mathbf{B}_1 = \begin{bmatrix} \sqrt{\frac{3}{4}} & & \\ & \sqrt{\frac{3}{4}} & \\ & & \sqrt{\frac{3}{2}} \end{bmatrix} \begin{bmatrix} \tilde{s}_1 & -\tilde{s}_2^* & \tilde{s}_5 & -\tilde{s}_6^* \\ \tilde{s}_2 & \tilde{s}_1^* & \tilde{s}_6 & \tilde{s}_5^* \\ \tilde{s}_7 & \tilde{s}_8^* & \tilde{s}_3 & -\tilde{s}_4^* \end{bmatrix}$$

$$\mathbf{B}_2 = \begin{bmatrix} 0 & 1 & 0 \\ 0 & 0 & 1 \\ 1 & 0 & 0 \end{bmatrix} \mathbf{B}_1$$

$$B_2 = \begin{bmatrix} 0 & 0 & 1 \\ 1 & 0 & 0 \\ 0 & 1 & 0 \end{bmatrix} B_1$$

$$C = \begin{bmatrix} s_1 \\ s_2 \\ s_3 \end{bmatrix}$$

- 4-antenna BTS downlink

$$A = \begin{bmatrix} s_1 & -s_2^* & 0 & 0 \\ s_2 & s_1^* & 0 & 0 \\ 0 & 0 & s_3 & -s_4^* \\ 0 & 0 & s_4 & s_3^* \end{bmatrix}$$

$$B = \begin{bmatrix} s_1 & -s_2^* & s_s & -s_7^* \\ s_2 & s_1^* & s_6 & -s_8^* \\ s_3 & -s_4^* & s_7 & -s_5^* \\ s_4 & s_3^* & s_8 & s_6^* \end{bmatrix}$$

$$C = \begin{bmatrix} s_1 \\ s_2 \\ s_3 \\ s_4 \end{bmatrix}$$

Similar to UL, the DL MIMO requires dedicated zones specified by the IE in DL-MAP. Readers are referred to the standard draft for the pilot patterns.

Access and ranging

Initial access, periodic synchronization, handoff, and bandwidth request are carried out by an MSS through ranging channels. Ranging is an essential means for the MSS to synchronize to the BTS for the first time.

A ranging channel is composed of one or more groups of six adjacent subchannels. Optionally, a ranging channel can be composed of eight adjacent subchannels. 256 binary ranging codes are grouped into four sub-groups, namely, (1) initial-ranging, (2) periodic-ranging, (3) bandwidth-requests, and (4) handover-ranging. Each user randomly chooses one ranging code from a band of specified binary codes. These codes are then BPSK modulated onto the subcarriers in the ranging channels – one bit per subcarrier. Collision may occur during the ranging process and multiple ranging signals can be separated using collision resolution algorithms.

The BS shall use the received ranging signal to estimate the timing offset and transmitting power of MSS so that proper adjustment is made before the MSS

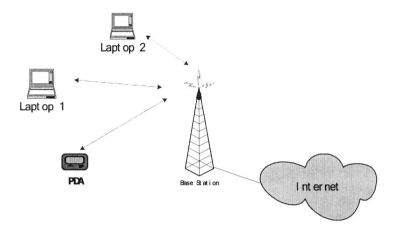

Figure 8.26: The PMP mode in 802.16 networks

is allowed to the network. The detailed responsibilities of the ranging channel processing at the BS include

- Detection: The BS needs to identify any MSS transmitting on the ranging channel. In the case that multiple MSSs are transmitting simultaneously on the same channel and cause a BS detection failure, a collision shall be declared by the BS.

- Timing offset estimation: The transmission timing offset of the MSS in reference to the BS timing needs to be estimated during the ranging process.

- Power measurement: The MSS transmits the ranging signal with the TX power level estimated from the open-loop power control. The BS shall measure the power of the received ranging signal to perform the closed-loop power control.

8.2.3 MAC layer technologies

Network architecture

The 802.16 MAC supports two types of network architecture: Point to Multi-Point network and Mesh network

Figure 8.26 illustrates a standard PMP network, where each terminal communicates with a BS. Terminals in 802.16 PMP network are termed subscriber stations (SS) and they do not communicate with each other directly.

Figure 8.27 illustrates an 802.16 Mesh network. Unlike the PMP mode, which is mandatory, Mesh mode is an option in 802.16. The main difference

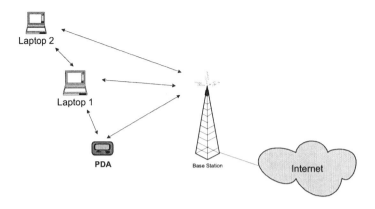

Figure 8.27: The Mesh mode in 802.16 networks

between PMP and Mesh modes is that in PMP mode, traffic only occurs between
BS and SSs, while in the Mesh mode traffic can be routed through other SSs
and can occur between SSs.

In general, a terminal in a Mesh network is termed a *node*. However, the
system that connects directly to backhaul service outside the Mesh network
is also termed a Mesh BS, while other terminals are termed Mesh SSs. The
downlink and uplink are defined as the traffic directed from/to the Mesh BS.

MAC sublayers and functions

The 802.16 MAC is composed of three sublayers as shown in Figure 8.28.

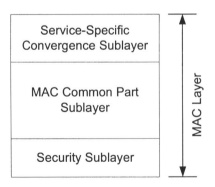

Figure 8.28: 802.16 MAC sublayers

1. Service-specific Convergence Sublayer (CS)

- The CS sublayer resides on top of the MAC Common Part Sublayer. In general, the CS sublayer distinguishes the type of higher layer packet data units (PDUs) and works with the appropriate MAC service access point (SAP) to deliver these PDUs. The CS also performs the following functions:

  - accepting PDU from the higher layer
  - performing classification and processing (if necessary) of higher layer PDUs
  - receiving CS PDU from the peer entity.

  At present, the 802.16 standard defines two CS specifications: the asynchronous transfer mode (ATM) CS and the IP-based packet CS.

2. MAC Common Part Sublayer (CPS)

- MAC CPS provides the core functionality of the MAC layer. It regulates the system access and connection establishment and management. The main MAC layer functions are summarized as

  - addressing and connection establishment and maintenance
  - MAC PDU construction, fragmentation and assembly
  - ARQ realization
  - scheduling services
  - bandwidth request and allocation
  - support of different PHY schemes
  - contention resolution
  - initial network entry and ranging
  - service flow based QoS provision
  - mobility support

3. Security Sublayer

- The security sublayer provides authentication, security key exchange and encryption. Currently, 802.16 security contains the following two component protocols:

  - an encapsulation protocol which encrypts the packet data during communication. The protocol defines the pairing of data encryption and authentication algorithms and the rules for applying these algorithms to MAC PDUs.
  - a key management protocol which provides the secure keying data distribution from BS to SS. Through this protocol, the BS and SS synchronize the keying data. The BS can also use this protocol to enforce conditional access to network services.

MAC CPS sublayer performs the main functionality of system access, bandwidth request and allocation, connection establishment and management, handover support, etc. In the following, we focus on some key aspects of 802.16 MAC CPS.

MAC common part sublayer

The 802.16 MAC is connection-oriented. All data communication occurs in the context of connections. A connection is associated with a service flow which defines the QoS parameters for the exchanged PDUs. Upon entering the network, two mandatory management connections and an optional connection are setup for the SS in both uplink and downlink. These three connections reflect the three levels of QoS requirements for management purposes. The first management connection is termed the *basic connection*, which is used to exchange short and delay sensitive management information; e.g., reset command and the burst profile request. The secondary connection is termed the *primary connection*, which is used to carry longer and more delay-tolerant information such as dynamic service configurations. The third connection is termed the *second connection* and is used to carry delay non-sensitive management information such as Dynamic Host Configuration Protocol (DHCP). Each connection is assigned a 16-bit connection identifier which can distinguish 64K connections within each downlink and uplink channel.

- MAC PDU format

  All MAC PDUs are constructed according to the format illustrated in Figure 8.29. A MAC PDU contains a fixed-length header and may or may not contain the Payload and CRC.

  The header fields are listed in Table 8.14

  Multiple MAC PDUs may be concatenated into a single transmission burst in either uplink or downlink to facilitate fast data transmission. One MAC SDU may be fragmented into multiple PDUs to efficiently utilize the available bandwidth relative to the QoS requirements. In addition, multiple MAC SDUs may be packed into one MAC PDU to allow efficient use of the bandwidth.

- Service Scheduling

  Each connection in 802.16 is associated with a data service and each data service is associated with a set of QoS parameters that reflect the service requirement. The 802.16 standard defines four types of services: Unsolicited Grant Service (UGS), Real-time Polling Service (rtPS), Non-real-time Polling Service (nrtPS) and Best Effort (BE). A brief exemplary description of the four services is illustrated in Table (8.10). The QoS parameters for the four services are listed in Table 8.15.

  Data transmission is scheduled by BS in downlink and by SS in uplink. The scheduler shall consider the following factors in determining the data transmission for a particular frame/bandwidth allocation:

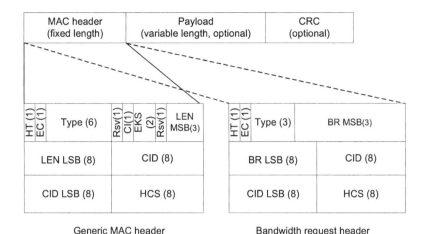

Figure 8.29: MAC PDU format

– the service specified for the service flow

– the values assigned to the service flow's QoS parameters

– the availability of data for transmission

– the capacity of the granted bandwidth

For uplink request/grant transmission scheduling, the 802.16 MAC defines the scheduling services and poll/grant options as shown in Table 8.16. However, the detailed scheduling algorithm is vendor-specific and the MAC only provides the means to implement these scheduling algorithms.

• Bandwidth Allocation and Request Mechanism

When an SS finds the allocated bandwidth insufficient, it needs to request bandwidth from the BS. The 802.16 MAC provides three types of mechanisms for the SS to send the bandwidth request message to the BS.

1. Request

Request is the mechanism by which the SS indicates to the BS that it needs uplink bandwidth allocation. The SS may send the request as a stand-alone bandwidth request header or PiggyBack the request in the Payload.

Bandwidth request messages may be incremental or aggregate. When the SS sends an incremental bandwidth request, the BS shall add the requested bandwidth quantity to the current bandwidth assigned to the connection. When the SS sends an aggregated bandwidth request, the BS shall store the requested bandwidth quantity as the effective bandwidth for the connection.

Field	Description
HT	Header type: 0=generic header; 1=bandwidth request (BR) header
EC	Encryption control: whether the payload is encrypted. BR header shall set the field to 0
Type	This field indicates the subheaders and special payload types in the message payload
Rsv	Reserved
CI	CRC indicator: whether CRC is included
EKS	Encryption Key Sequence: the parameter to encrypt the payload
LEN	The length in bytes of the MAC PDU including MAC header and CRC (if present)
CID	Connection ID
HCS	Header Check Sequence: a sequence calcuated according to a pre-defined formula
BR	Bandwidth Request: the number of bytes request for uplink by the

Table 8.14: MAC PDU header fields

2. Grants

   When an SS sends the bandwidth request, it is associated with an individual connection, while each bandwidth grant is always addressed to the SS's basic connection, not the requested connection. As a result, when the SS receives a shorter transmission opportunity than expected, it does not know for which connection the bandwidth request is not granted. The SS shall decide whether to retransmit again or discard the request.

3. Polling

   Polling is the mechanism by which the BS allocates the bandwidth to an individual SS or a group of SSs specifically for the purpose of making bandwidth requests. Two types of polling are defined: unicast polling and multicast/broadcast polling.

- Mobility Support

  Due to mobility or BS load consideration, an SS may need to change connection in order to obtain better signal quality. A handover begins with the decision for an MS to hand-over its air interface, service flows and network attachment from a serving BS to a target BS. The handover may be initiated by the MS, the BS or from the network side. The 802.16e handover process consists of the following steps illustrated in Figure 8.30: 1) cell-reselection, 2) handover decision and initiation, 3) target BS scanning, 4) network re-entry and 5) termination of service.

Service Type	QoS Parameters
UGS	Maximum Sustained Traffic Rate, Maximum Latency, Tolerated Jitter and Request/Transmission Policy
rtPS	Minimum Reserved Traffic Rate, Maximum Sustained Traffic Rate, Maximum Latency and Request/Transmission Policy
nrtPS	Minimum Reserved Traffic Rate, Maximum Sustained Traffic Rate, Traffic Priority and Request/Transmission Policy
BE	Maximum Sustained Traffic Rate, Traffic Priority and Request/Transmission Policy

Table 8.15: Service QoS parameters

Service Type	PiggyBack Request	Bandwidth Stealing	Polling
UGS	Not allowed	Not allowed	Use a special bit to request a unitcast polls for non-UGS connections
rtPS	Allowed	Allowed	Only allowd unitcase polling
nrtPS	Allowed	Allowed	May restric a service flow to unicast polling via the transmission/request policy; otherwise all forms of polling is allowd
BE	Allowed	Allowed	All forms of polling allowed

Table 8.16: Service scheduling options

8.3 Performance analysis of WiMAX systems

Characterizing the performance of a radio frequency telecommunications system poses interesting challenges. Applying modulation rates to channel bandwidth only yields a per sector data rate which is an interesting figure but not sufficient. To fully understand the performance of a radio frequency telecommunications system it is vital to include (1) sector data rates, (2) the service levels sold, and (3) the sector loading efficiencies allowing for perceived broadband data rates to answer the question of system performance which culminates in the metric of base station *capital expense per megabit per second per subscriber* (capex / mbps /subscriber). Following is an analysis of a 5Mhz OFDMA-TDD system according to this method.

8.3.1 WiMAX OFDMA-TDD

OFDMA is the air interface adopted by the worldwide WiMAX and Korean WiBro standards as the technology for mobile broadband connectivity. OFDMA

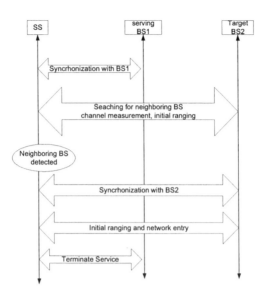

Figure 8.30: Handover process

uses the multi-channel OFDM approach and provides subscriber access in the time domain (TDMA) and in the frequency domain (FDMA) and duplexes in time (TDD). Decisions as to which timeslot, subchannel, and power level to communicate over are determined by the intelligent MAC which seeks to maximize the SINR for every subscriber. This allows subscribers to operate at the maximum modulation rates obtainable given the radio frequency conditions at the subscriber location. This allows service providers to maximize the number of active subscribers in each sector whether the subscriber is fixed, portable, or mobile. The scalable OFDMA in IEEE 802.16e allows for software scalable FFT sizes to match the service provider's bandwidth allocation and maintain a proper subcarrier bandwidth providing for both multipath and Doppler immunity.

Time Division Duplexed (TDD) system offers flexibility in configuring the downlink:uplink ratio to match the subscriber service levels and modulation rates. This results in a superior use of spectrum over Frequency Division Duplexed (FDD) systems which cannot adjust for asymmetric data traffic and the differences in uplink and downlink modulation rates and tend to be unbalanced in uplink and downlink capacity.

8.3.2 Comparison method

In order to quantify the performance of the ADAPTIX system we have included the perspectives of the (1) sector data rates, (2) the service levels sold, and (3) the sector loading efficiencies allowing for perceived broadband data rates to answer the question of system performance which culminates in the metric of base station capital expense per megabit per second per subscriber (capex / mbps /subscriber). To accomplish this the following analyses are made:

1. Establish service levels and overbooking

2. Determine perceived and committed data rates

3. Determine per sector data rates at maximum modulation

4. Determine per sector data rates at blended modulation

5. Calculate subscribers at 100% sector loading

6. Calculate sector loading efficiency

7. Calculate subscribers at sector loading efficiency

8. Calculate capex / mbps / subscriber

Establish service levels and overbooking

The system performance analysis is started by assuming service level offerings and corresponding overbooking factors. Overbooking defines the number of subscribers accessing the network at a statistical level such that each of those subscribers has the perception of having that data connection to themselves.

1. Establish service levels and overbooking
overbooking = number of subscribers sharing a connection, each perceiving to have it to themselves

service level	overbooking
64	10
144	30
384	25
512	20
768	20
1000	15
1500	10
2500	5

Determine perceived and committed data rates

The perceived blended data rate can then be calculated by applying the distribution to the respective service levels. The committed data rates can be determined by dividing the various service levels by their respective overbooking factors and the weighted average yields the blended committed data rate.

2. Determine perceived and committed data rates

kbps / service = service level / overbooking factor

distribution	DL	av kbps/sub	kbps/sub	UL	av kbps/sub	kbps/sub
20%	64	6	1.28	64	6	1.28
20%	512	26	5.12	384	15	3.07
20%	768	38	7.68	512	26	5.12
20%	1000	67	13.33	512	26	5.12
20%	1500	150	30.00	768	38	7.68
wt averages	768.8		57.41	448		22.27

The blended downlink service level is 768kbps with a corresponding committed data rate of 57kbps, and the blended uplink service level is 448kbps with a corresponding committed data rate of 22kbps.

Determine maximum per sector data rates

The performance of radio communications systems varies greatly depending on factors that include system power, antenna gain, Rayleigh fading, Doppler, multipath, ambient RF noise, and propagation effects. It is, however, still informative to assume the best conditions and determine the system throughput. These figures also support burst rate assertions assuming there is only one active subscriber in the sector operating at the best modulation rates.

3. Determine sector data rates at maximum modulation rates

TCP/IP header + data + checksum only, marketing numbers

TDD Ratio	1.35 :1			

Bandwidth (TDD)	5			
	dl %	DL	UL	ul %
64QAM 5/6	100%	12.77	9.46	0%
64QAM 2/3	0%	10.21	7.57	0%
16QAM 3/4	0%	7.66	5.67	100%
16QAM 1/2	0%	5.11	3.78	0%
QPSK 3/4	0%	3.83	2.84	0%
QPSK 1/2	0%	2.55	1.89	0%
raw data rates	100%	12.77	5.67	100%
minus overhead	-20%	10.15	4.51	
bits per hz		2.96	2.88	2.93

It is important to note that in these analyses we are only interested in the data rates that correspond to a salable service, not the raw data rates that include the overhead associated with the radio system. The overhead figure in this case is 20% of the raw data rate but results in a salable throughput data rate of:

Salable data rate=TCP/IP header + data + checksum

Determine per sector data rates at blended modulation

Now that we have determined the maximum performance it is time to establish the probable performance using blended modulation rates.

The OFDMA modulation rates can be determined by a COST231 propagation model which is a first order approximation for radio communications system sizing. Given these service levels the 5Mhz OFDMA-TDD system provides data rates of 6.52Mbps downlink and 2.68Mbps uplink.

4. Determine sector data rates at blended modulation rates

TCP/IP header + data + checksum data rates, real world deployment propagation characteristics

| TDD Ratio | 1.35 :1 | | | |

Bandwidth (TDD)	5			
	dl %	DL	UL	ul %
64QAM 5/6	15%	12.77	9.46	0%
64QAM 2/3	26%	10.21	7.57	0%
16QAM 3/4	25%	7.66	5.67	19%
16QAM 1/2	35%	5.11	3.78	26%
QPSK 3/4	0%	3.83	2.84	28%
QPSK 1/2	0%	2.55	1.89	27%
raw data rates	100%	8.19	3.36	100%
minus overhead	-20%	6.52	2.68	
bits per hz		2.27	1.26	1.84

Calculate subscribers at 100% sector loading

Then by dividing the downlink and uplink sector data rates by the blended committed data rate the number of users at 100% system capacity can be calculated.

5. Calculate number of subscribers at 100% sector loading

number of subscribers / sector = sector data rate / committed data rate

DL	UL
113.51	120.14

The TDD systems can balance the downlink and uplink symmetry to match the service level and subscriber usage pattern and therefore make better use of the available spectrum.

User Perception of Broadband Service of a Shared Resource

A shared resource such as a sector of a radio system or a cable system supplying data connectivity takes the bandwidth of the system and divides it among the subscribers to that service. Three studies done by N.K. Shankaranarayanan et al at AT&T Labs from 2001 to 2003 explain the issues surrounding perceived data throughput in a shared resource environment [8][9]. He has determined that in order for megabit per second subscriber experience two criteria must be met:

1. The committed data rates must be at least 35kbps

2. The network utilization is a function of perceived and system data rates

Calculate sector loading efficiency

This relationship between the perceived data rate and system data rate is important in sizing shared resource networks such as cable or radio sector based. The notion behind this relationship is if the network were at 100% capacity then a subscriber in an idle mode (reading downloaded material) transitioning to an active mode (requesting the next bit of information) would have to wait in the queue until there is an opening for that request to be serviced. This queue period degrades the perception of megabit per second broadband connectivity. In order to size the utilization factor of this shared resource (wireless communications sector) N.K. Shankaranarayanan [8][9] has established the following relationship:

$$PDR=SDR \times (1\text{-sector loading efficiency})$$

where the PDR is the Perceived Data Rate and the SDR is the Sector Data Rate.

6. Calculate sector loading efficiency to support perceived data rate
pdr = sdr x (1 - sector loading efficiency)   formulated by NK Shankaranarayanan at ATT Labs

DL	UL
88.20%	83.26%

As is evident by this relationship, narrowband systems with low sector data rates trying to provide megabit per second services must run at low sector loading efficiencies.

Calculate subscribers at sector loading efficiency

By applying the sector loading efficiencies to the number of subscribers at 100% capacity the number of serviceable subscribers can be calculated. By applying the distribution of the service levels against the number of serviceable subscribers the number of subscribers at each service level can be calculated.

7. Calculate number of subscribers at sector loading efficiencies

number of service supportable subscribers = subscribers at 100% x respective sector loading efficiency

	DL	UL	min(DL, UL)
	100.11	100.02	100.02
distribution	DL	UL	subscribers
20%	64	64	20
20%	512	384	20
20%	768	512	20
20%	1000	512	20
20%	1500	768	20

Calculate capex / mbps / subscriber

In order to quantify the performance of the OFDMA-TDD system we include the perspectives of the (1) sector data rates, (2) the service levels sold, and (3) the sector loading efficiencies allowing for perceived broadband data rates to answer the question of system performance which culminates in the metric of base station capital expense per megabit per second per subscriber (capex / mbps /subscriber).

This comparison metric displays the advantages of broadband channels using high modulation rates resulting in high per sector data rates and large number of supported active subscribers.

8. Calculate capex / mbps / subscriber

capex / mbps / subscriber = base station capex / sector mbps / number of subscribers

base station capex	$30,000
mbps / sector	9.19
subscribers / sector	100
$ / mbps / sub	$33

In summary, an OFDMA-TDD system is:

- extremely spectrally efficient;

- able to operate high sector loading efficiencies;

- able to support many active subscribers with high data rates;

- resulting in low sector capital cost per Mbps per subscriber.

Bibliography

[1] IEEE Std 802.11-1997 IEEE Std 802.11-1997 Information Technology-Telecommunications and Information Exchange Between Systems-Local and Metropolitan Area Networks-Specific Requirements–part 11: Wireless LAN Medium Access Control (MAC) and Physical Layer (PHY) Specifications.

[2] IEEE Std 802.11a-1999 IEEE Std 802.11a-1999 Information Technology-Telecommunications and Information Exchange Between Systems-Local and Metropolitan Area Networks-Specific Requirements–Part 11: Wireless LAN Medium Access Control (MAC) and Physical Layer (PHY) Specifications: High-speed Physical Layer in the 5 GHz Band.

[3] IEEE Std 802.11b-1999 IEEE Std 802.11a-1999 Information Technology-Telecommunications and Information Exchange Between Systems-Local and Metropolitan Area Networks-Specific Requirements–Part 11: Wireless Lan Medium Access Control (MAC) and Physical Layer (PHY) Specifications: High-speed Physical Layer Extension in the 2.4 GHz Band.

[4] IEEE Std 802.11g-2003 IEEE Std 802.11a-2003 Information Technology-Telecommunications and Information Exchange Between Systems-Local and Metropolitan Area Networks-Specific Requirements–Part 11: Wireless LAN Medium Access Control (MAC) and Physical Layer (PHY) Specifications: Further Higher Data Rate Extension in the 2.4 GHz Band.

[5] ETS 300 744, "Digital video broadcasting; framing. structure, channel coding and modulation for digital terrestrial television (DVB-T)." European Telecommunications Standards Institute ETSI, January 1999.

[6] IEEE P902.16-2004, "Standard for local and metropolitan area networks Part 16: air interface for fixed broadband wireless access systems," URL: Http://Grouper.ieee.org/groups/802/16/tgd/.

[7] H. Yaghoobi, "Scalable OFDMA physical layer in IEEE 802.16 Wireless-MAN," *Intel Technology Journal*, vol. 8, issue 3, 2004.

[8] Z. Jiang and N. K. Shankaranarayanan, "Channel Quality Dependent Scheduling for Flexible Wireless Resource Control," *Proc Globecom 2001*, San Antonio, TX, Nov. 2001.

[9] N. K. Shankaranarayanan, Anupam Rastogi, and Zhimei Jiang, "Performance of a wireless data network with mixed interactive user workloads", *Proc. IEEE Int. Conf. Commun., ICC'02*, New York, April 2002.

Notations, Acronyms and Commonly Used Symbols

Notations

x	the scalar x
$\mathbf{x}$	the vector $\mathbf{x}$
$\arg\max_k (x_1, x_2, ..)$	the index of the maximum value over $x_1, x_2, ..$
$\arg\min_k (x_1, x_2, ..)$	the index of the minimum value over $x_1, x_2, ..$
$O(\cdot)$	on the order of $\cdot$
$dom\ f$	the domain of function f, i.e., $\{x \mid f(x) < +\infty\}$
$\odot$	the element-wise (Schur-Hadamard) product
$*$	convolution
R_+	the nonnegative real space
$\|\cdot\|$	the Euclidean norm for vectors
$E\{\}$	the expectation of random variables
$diag(\mathbf{x})$	the square matrix whose diagonal elements are $\mathbf{x}$
$(\cdot)^H$	the conjugate transpose
$(\cdot)^T$	the transpose
$Re\{\cdot\}$	the real part
$Im\{\cdot\}$	the imaginary part
$(x)^+$	the maximum value between 0 and x
$\frac{df}{dx}(w)$	the first derivative of function $f(x)$ evaluated at w
$\nabla f(\mathbf{w})$	the gradient of function $f(\mathbf{x})$ evaluated at $\mathbf{w}$

Acronyms

ACI Adjacent Channel Interference

ACM Adaptive Coding and Modulation

AP Access Point

AWGN Additive White Gaussian Noise

BER Bit Error Rate

BPSK Binary Phase Shift Keying

BS Base Station

BSS Basic Service Set

CDF Cumulative Distribution Function

CDMA Code Division Multiple Access

CP Cyclic Prefix

CPE Common Phase Error

CSI Channel State Information

CSMA/CA Carrier Sensing Medium Access/Collision Avoidance

DCA Dynamic Channel Allocation

DMT Discret MultiTone

DSL Digital Subscriber Line

DSSS Direct Sequence Spread Spectrum

DVB-T Digital Video Broadcasting – Terrestrial

ESS Extended Service Set

FCA Fixed Channel Allocation

FDD Frequency Division Duplexing

FDMA Frequency Division Multiple Access

FHSS Frequency Hopping Spread Spectrum

FIR Finite Impulse Response

GSM Global System for Mobile communication

H-ARQ Hybrid Automatic Repeat reQuest

HCA Hybrid Channel Allocation

HDTV High-Definition Television

IBSS Independent Basic Service Set

ICI Inter-Channel Interference

IIR Infinite Impulse Response

ISI Inter-Symbol Interference

ISM Industrial Scientific and Medical

LAN Local Area Network

LTV Linear Time Variant

MAC Multiple Access Control or Medium Access Control

MAN Metropolitan Area Network

MC-CDMA Multicarrier CDMA

MIMO Multiple Input Multiple Output

MMSE Minimum Mean Square Error

OFDM Orthogonal Frequency Multiplexing Division

OFDMA Orthogonal Frequency Division Multiple Access

OSTBC Orthogonal Space-Time code

PAPR Peak to Average Power Ratio

pdf Probability Density Function

PLL Phase Lock Loop

PHY Physical (Layer)

QAM Quadrature Amplitude Modulation

QoS Quality of Service

QPSK Quadrature Phase Shift Keying

RNC Radio Network Controller

SDMA Spatial Division Multiple Access

SINR Signal to Interference and Noise Ratio

SISO Single Input Single Output

SNR Signal to Noise Ratio

STC Space-Time Code

TCP Transmission Control Protocol

TDD Time Division Duplex

TDMA Time Division Multiple Access

UNII Unlicensed National Information Infrastructure

WiFi Wireless Fidelity

WiMAX Wireless (Wi) Microwave Access (MA)

ZF Zero Forcing

Commonly Used Symbols

p	transmission power
h	channel gain scaler
$\mathbf{h}$	channel gain vector
$\mathbf{H}$	channel gain matrix
N_0	AWGN spectral density
N_t	the number of transmit antennas in MIMO systems
N_r	the number of receive antennas in MIMO systems
$\psi_x(j\omega)$	the characteristic function of random variable x
$\{\lambda_i\}$	the eigenvalues of a matrix
$\mathbf{I}$	identify matrix
Q	total transmission power
q_n	total transmission power on subcarrier n

About the Authors

Dr. Hui Liu is one of the premier authorities on broadband wireless technology, with a proven record of turning concepts into real systems. He is currently an associate professor at the Department of Electrical Engineering, University of Washington, Seattle. As the director of Wireless Information Technology (WIT) Lab, he supervises a wide range of research from broadband wireless networks, to DSP and VLSI for communications, to multimedia signal processing. Previously Dr. Liu was the chief scientist at Cwill Telecommunications, Inc., and was one of the principal developers of the TD-SCDMA technologies. He founded Broadstorm Inc., in 2000, and was responsible for the overall architecture of the company's highly innovative OFDMA mobile broadband Internet solution.

As an author, Dr. Liu has 40 published journal articles, as well as the book *Signal Processing Applications in CDMA Communications*, published by Artech House Publishers in 2000. Dr. Liu is active in the IEEE Communications Society, including membership on several technical committees and serving as an editor for IEEE Transactions on Communications. He is the general chairman for the 2005 Asilomar Conference on Signals, Systems, and Computers. He is a recipient of 1997 National Science Foundation (NSF) CAREER Award and the 2000 Office of Naval Research (ONR) Young Investigator Award. One of his patents, *Smart Antenna TDD CDMA Wireless Communication Systems*, won the China National Gold Medal Award. He also received the Best Paper Award in the 2003 FPL conference and supervised students with awarding-winning student papers.

Dr. Liu received a B.S. in 1988 from Fudan University, Shanghai, China; an M.S. in 1992 from Portland State University, Portland, Oregon; and a Ph.D. degree in 1995 from the University of Texas at Austin, all in electrical engineering. He held the position of assistant professor at the Department of Electrical Engineering at University of Virginia from September 1995 to July 1998.

Dr. Guoqing Li received a B.S and an M.S. from the University of Science and Technology of China, Hefei, China, in 1998 and 2001, respectively. She received a Ph.D. in Electrical Engineering from University of Washington, Seattle, in 2004. Prior to her Ph.D. study, Dr. Li had worked as a senior system engineer in UTStarcom, Inc., developing algorithms for physical, multiple access control, radio link control and radio resource control layers for 3G-WCDMA and IS-95 networks. She also has worked on GSM/GPRS/EDGE technologies in T-Mobile USA, where she built the GSM/GPRS handset roadmap and features enhancements. Dr. Guoqing Li has published over 20 journal and conference papers, authored a patent and received numerous awards, including the 2004 Asilomar conference student paper awards and the Outstanding Graduate Award. Dr. Li is a research scientist in Intel Corporation, Communication Technology Lab. Her research interests include broadband wireless networks, OFDM/OFDMA technologies, resource management in wireless communication systems and cross-layer design in mesh networks.

Index